THE STORY BEHIND THE FOUNDERS OF THE ESCAPEES RV CLUB

Each of us has a story.
This is mine.
I hope you enjoy it.
Kay Peterson

Kay Peterson

outskirtspress
DENVER, COLORADO

Some names have been changed for privacy reasons. It is not my intention to hurt or judge anyone, but rather to show how our interaction impacted either Joe's or my life. Conversations have been recreated from my memory to make our story more readable. All events are portrayed from my memory of them.

Beating the Odds
The Story Behind the Founders of the Escapees RV Club

v3.0

Outskirts Press, Inc.
http://www.outskirtspress.com

Paperback ISBN: 978-1-4327-9850-5
Hardback ISBN: 978-1-4787-2080-5

PRINTED IN THE UNITED STATES OF AMERICA

This book is dedicated to Joe, of course.

Acknowledgements:

MY THANKS TO editor and friend, Susie Gearing, to Outskirts Publishing, and to Cathie for their technical help.

Contents

Prelude: Family History

MY PATERNAL GREAT-GRANDFATHER was a 12-year-old orphan in 1834 when England passed the "poor law" forcing orphans and the homeless to emigrate to one of England's colonies. My great-grandfather was indentured to a Canadian farmer, who paid his passage in return for his work, until he reached age 18. Then he received his own land where he built the cabin he lived in until he died at age 84.

He was a self-taught musician, made his own fiddle and played at military dances. There he met the daughter of an army officer. They fell in love, but her father refused to let her marry someone with no family background. When she learned her parents were sending her to England, she ran away, married my great-grandfather, and they had four children.

Under English law, my grandfather inherited the homestead, married Mel, and they had 12 children. He worked in the lumber camp in winter to subsidize farming. In 1914, at age 52, a tree fell on him. He became an invalid and died a year later.

Mel's two oldest boys were drafted into Canada's army during World War I, leaving her with a homestead that could not support her remaining family. She found a job as housekeeper/cook but could only take her youngest girl with her. The rest of the children had to find a way to support themselves. The girls got married or worked as domestics.

My father was the next youngest child. He had just turned 10 and loved school. That life was over. If he wanted to eat, he had to work. He and his 12-year-old brother stayed on the homestead that summer, working in the lumber mill, stacking wood to dry.

"The piles were taller than I was," Dad said. "I'd cry when I had to restack a pile that fell down."

In fall, the potato crop was harvested. Men dug up the potatoes and left them lying in the fields for boys to collect. The boys were paid 25 cents for a 10-hour day. After that job ended, Dad and his brother went to a lumber camp where men felled huge trees, using only an axe and crosscut saw. His brother helped pile logs, but Dad worked in the camp kitchen, helping prepare four meals a day. Everyone worked from sunup till sundown with no days off. This was Dad's life for eight years. When he turned 18, he migrated to the U.S. and became a citizen a year later.

He was a small man, 5′8″, weighed 150 pounds, and had health problems. He went to Massachusetts where his older brothers had settled after World War I ended. He drove a milk delivery truck, worked as a short-order cook in a diner, and later delivered groceries. While working at the diner, he met my mother, a petite auburn-haired telephone operator who ate lunch there.

My other grandfather, Sidney, also an Englishman, was sent to Maine to teach spinning-mill techniques. There he met and married Mabel. It was not a happy marriage partly because of religious differences. When his job ended, Sidney wanted his family to go to England, but Mabel refused. It was a bitter ending to their marriage and something she would never talk about.

My mother, Betty, was sent to a Catholic boarding school. She was 13 when her father died. Then Mabel remarried, and my mother went to live with them until she was 18. She moved to Massachusetts and went to work as a telephone operator. She was less than five foot and looked as if a strong wind would blow her over. She proved tougher than she looked.

My parents met, fell in love and married in 1925. They had three children: Alice born in 1926, Catherine (me) in 1927, and Roy in 1929. Roy had a bad heart. Today, it could have been repaired with surgery enabling him to live a normal life, but in the 1930s that technology didn't exist. If it had, we could not have afforded it. There was no health insurance; people paid for hospital care or went without.

A family doctor made home visits to his patients even if he knew he would not get paid. He charged what he thought the family could afford. We also used home cures passed down through generations. Some worked. Others, like burying a hen's head under your front stoop, were a last and useless resort.

After Roy was born, Mama developed ovarian cancer. It spread to one breast and then the other. People didn't understand cancer, so they feared it; victims were ashamed of having it. Mama believed losing her breasts took her womanhood. She died still thinking she was a failure as a woman.

Cancer research was in its infancy, and Mama qualified for a research program on radiation treatment. It paid for her hospitalization and multiple surgeries, which Dad never could have afforded. The downside was a research team decided when to operate and how much radiation to give. (Chemotherapy did not exist.) They were still experimenting with radiation dosages. Some doctors believed it was extreme radiation that allowed her to live with cancer for over 17 years, but the treatments made her extremely ill. I never heard her complain.

She was a victim of cancer for the rest of their marriage. Between cancer treatments, she worked as a maid in a hotel. When cancer spread to her head causing a lump under one eye that looked as if someone hit her, she asked to be moved to the laundry room. It meant standing on her feet ironing sheets and pillow cases all day, but she no longer had to put up with embarrassing questions or judgmental stares.

Dad was able to get a job delivering groceries for a grocery store. He kept that job through most of my early childhood until the Depression forced the store to close its doors. We didn't have money to treat his anemia, so every morning on his way to work, Dad stopped at a slaughterhouse to get a free pint of fresh cow blood to drink.

Being without an education, finding temporary jobs to provide food and shelter during the Depression, and having to take care of a sick wife for weeks at a time showed what a high sense of responsibility and loyalty Dad had. What might he have achieved with just a few more years of education?

This is the background I came from. I believe it was good genes and being taught responsibility that enabled me to get through rough periods in my life and, eventually, achieve my dreams.

CHAPTER 1

Childhood Innocence (1927—1944)

MY EARLIEST MEMORY was waking up in a white crib with white sheets in a room with white walls. My throat hurt. A lot. I remembered being wheeled into a room—not this one—to have my tonsils taken out. They said my throat would hurt, but nobody said it would hurt this much. I started to cry, but that made it hurt worse.

I sat up. There were many cribs like mine occupied by children. Some were crying; others sleeping. A little girl stared at me through the bars and whimpered. I started to tell her everything would be all right, but when I tried to speak my throat caught on fire. I was scared. Would I ever be able to talk again?

Then an angel appeared and asked if I wanted ice cream. She didn't have wings like pictures of angels do, but she was dressed all in white and had a white hat on. When I nodded, she patted my hand. I grabbed hers, not wanting her to go. She didn't pull her hand away. With her free hand, she rubbed my hands that had captured hers.

"You'll feel better soon. I'll get your ice cream."

I let her go. From that day on, I knew I *had* to be a nurse. I had no idea how difficult fulfilling that dream would be.

I didn't realize we were poor. Everyone struggled. That was the way life was. Our clothes were always clean and ironed. Roy and Alice got their new clothes from the church charity store. My new clothes were my sister's hand-me-downs. It was exciting when anyone got a new outfit, no matter where it came from or who it was for.

We entertained ourselves without expensive toys. We played tag, dodge ball, hide-and-seek, cops and robbers, hopscotch, or skipped rope. In inclement weather, I played with both real and paper dolls. Each Sunday newspaper had a new paper doll or clothes from the previous one to be cut out. I played house with paper dolls and hospital using real dolls and my brother, when I could talk him into it, as patients.

Newspapers and radio were our communication with the outside world. We gathered around the radio, listening to whatever program was on. My favorite was "The Shadow" with its sinister laugh and squeaking door trademark.

Our other family entertainment was visiting our grandparents who owned a chicken farm. We fed chickens and gathered eggs to take to the cellar where Grandma placed each on a scale and then told us which crate to put it in. Her scale had a light so she could see if there was a blood spot inside. Eggs that were cracked or had a blood spot were not sold. The family ate them.

We lived in cheap apartments. My worst memories were cleaning out an army of cockroaches each time we moved. No matter what you did, they survived. Many nights I woke up crying with bites on my body. Mama would tear the bed apart and spray the mattress and bedsprings with something to kill bed bugs.

I only remember getting into trouble twice as a child. The first happened in third grade and involved my favorite teacher, Miss Stratton. I was *almost eight*. Roy had started school, and I felt responsible for him although I do not know why. It was a burden of my own choosing.

A bully in my class threw a snowball hitting Roy in the face. I grabbed a chunk of ice and hurled it at the bully, hitting him under his eye. Wailing and holding his bleeding cheek, he ran to school. I knew he would tell Miss Stratton. I took as long as possible removing and hanging up my snow clothes.

Would she send me to the principal's office? I didn't know what happened there, but everyone returned with reddened eyes, a sign of crying. Miss Stratton called me to her desk. Placing both hands on my cheeks, she forced me to look into her eyes. "A chunk of ice is not a snowball. If it had hit an inch higher he could have lost that eye."

I didn't understand how getting hit in the eye could make you lose it. Maybe it sunk back inside your head leaving a big hole there. Much as I hated him, I didn't want him to walk around forever with a hole instead of an eye. I started crying.

Miss Stratton said, "I know you didn't mean to hurt him, but you *must think about what you are doing before you act.*" I still remember her advice, and it is when I don't follow it that I regret it.

In 1938, the war starting in Europe didn't affect Americans. In 1939, as I entered the eighth grade, Germany attacked Poland and then Great Britain. Several countries declared war on Germany, but Americans insisted it was not our concern. Politicians knew they would not be re-elected if they voted to enter a foreign war.

However, Congress started the draft, and men between 21 and 26 had to register. Men who still couldn't find good jobs began joining one of the services. This put a small crack in the job market for men with no education.

Dad went to work as a night watchman—his first steady job since the Depression.

Having a steady job allowed us to move to North Cambridge that summer between elementary and high school. We lived in the same apartment in this less populated part of the city throughout my high school years. It is the only house or apartment I remember.

There were only three apartment buildings plus a single house where the landlord lived. Each building had three floors with an apartment on each. One girl, older than we, lived in the first building but never spoke to us. We lived in the third-floor apartment of the second building, and a Mexican boy, Toto, lived in the third building. The rest were adults. Toto was a year younger than I and never knew I had a crush on him. Roy, Toto, and I mostly played together, except Alice played baseball with us.

The only other building in sight was a large brick building on a small hill across the street. It was "The Poorhouse." In those days, the hobos created by the Depression and people too old to work did not live on the street. Some said the poorhouse was a shameful place to live, but residents had a bed and three meals of some kind every day. What *is* shameful is that today's homeless sleep under bridges and behind back-alley trash bins, which they raid for rejected food.

The people in charge allowed us to play baseball in a field behind the poorhouse. Old men hobbled out to watch us and then sat with us, telling stories of lives lost through bad choices or the Depression. They were good people with lives ending sadly. Getting to know them inspired one of my dreams. Like seeds planted in spring, I harvested the results many years later.

We were also allowed to play on the lawn in front of the poorhouse. We wore bathing suits and chased each other through sprinklers while the old folks sat on their front porch and watched. It was here that I first forgot Miss Stratton's advice.

I saw Toto bent over the sprinkler getting a drink. To get his attention, I pushed his head down as I ran past. I wanted him to chase me and hold my head in the water for revenge, but the fun stopped when his front tooth hit the sprinkler head and broke off.

I repeatedly told him I was sorry. Being sorry doesn't help when your family can't afford a dentist. I wonder how long he wore the badge of my thoughtlessness. I know that throughout high school it was a constant reminder to me of Miss Stratton's advice.

I had just started high school when in November 1940 the impossible happened. Franklin Delano Roosevelt won a *third term*–the only President to run for a third term. *He won by a landslide* because the world was in chaos, because people feared the future, because Roosevelt introduced programs to give people jobs, and because we believed he kept us out of war.

Because I had started first grade at five instead of six, I was always younger than most classmates. Now at 13, my big concern was what was happening to my body. When I started my menstrual cycle, my sister showed me how to use the belt and pads Mama had bought for us. I wondered why girls had "periods" and boys didn't, but *no one* talked about it. Sex was a secret you found the answer to after you got married.

I cannot remember having any close friends except Toto. I made my first friends in high school. Except for those going to college, boys went to the trade school across the street. This meant there was a shortage of boys at my school.

My friends were the two girls in most of my classes. Louise was a year older and very popular. I wondered why she sometimes included me in her large circle of friends. We both chose nursing for a career, but except for that we had little in common. I felt closer to Maggie because we both were "outsiders." We ate lunch together, confiding our crushes to each other.

In February, I turned 14. Until then, the only way to earn money was babysitting. The pay was 50 cents *for the evening*, but if I washed their dishes and picked up the kids' clutter, they might give me 75 cents. I used this money to go bowling with the church youth group, to a Saturday afternoon movie (ten cents for a double feature), or to the roller skating rink.

Now I could apply for a real job and start saving for nurses training. Jobs were available because unskilled workers were now building ships and planes even though America was still not at war. I applied for a job at our community hospital and was hired for the women's ward kitchen, where I worked throughout my remaining high school years.

That same year, on December seventh, the Japanese bombed Pearl Harbor, destroying part of our navy. Everything changed. For me, life was better. The war opened that closed door for Dad. He went to work in the shipping department of Folgers Coffee company in Boston and worked there until he was forced to retire because of severe emphysema.

I was 15 and in my junior year when Louise, who was turning 17, invited me to her birthday party. I had never been to a birthday party except for relatives. When I asked if Maggie was invited, Louise said, "She's too young." Actually, Maggie was already 16, and I wouldn't be 16 until the end of February.

When I asked for permission to go, Mama wanted to know who would be there. I didn't know. She asked if Louise's parents would be there. I thought all parents are at home when their child has a party, but I promised to ask. When I did, Louise said, "Why do you ask?"

"It's my mother," I said, embarrassed. "She wanted to be sure your parents would be there."

I didn't realize how noncommittal her answer was. "Tell your mother not to worry." Taking it as a yes, I reassured Mama the parents would be there. Reluctantly, she agreed provided I come home on the 11 o'clock bus.

When I arrived, the party had already started. Louise's parents were nowhere in sight. Music was playing; people were talking in groups, drinking punch, and eating crackers with cheese. While I was pouring a glass of punch, a girl whispered, "Take it easy. It's spiked."

I didn't know what she meant, but it tasted awful, so I only took a sip when someone looked at me, which was not often. The few people I recognized were not my friends. I missed Maggie.

Then I saw the football star, a handsome popular senior whose name I have long since forgotten. Like many others, I admired him from afar and fantasized what it would be like to be "his girl." I sat in a chair and watched how easily people talked and switched groups. How come I couldn't do that? I tried to tell myself it was because I didn't know them, but I knew it was more than that.

Someone suggested playing spin-the-bottle. I was relieved to have something to do. A circle was forming with people sitting cross-legged on the floor. When I saw the football star was part of that circle, I eagerly joined them. I felt even better when Louise sat beside me. I would do whatever she did.

When the game started, I realized it was a kissing game! My heart began pounding. How do you kiss a boy? In the movies, two people gaze into each other's eyes and then their lips meet. The scene immediately shifts to a curtain blowing in the wind, so you never saw the ending of the kiss. Now I would see how the kiss ended.

I was disappointed. No one gazed into each other's eyes. They locked lips immediately and then it was over. They returned to their seats with a gloating smile. Hey, I could do that!

After several people had their turn spinning the bottle and claiming their kisses, the bottle came to rest between Louise and me. Without hesitation, the boy announced it was pointing at Louise, and she jumped up to claim her kiss. I felt a surge of relief. I didn't want to kiss a boy I didn't even know. I hoped it wouldn't stop in front of me.

There were many more spins, during which the kisses became more prolonged. Then the football star had a turn spinning the bottle. This time I *wanted* it to land in front of me. I think I willed it to do so. When it came to rest squarely in front of me, I felt my face flushing while my mind rejoiced. My reaction was short-lived.

Louise pushed the bottle so that it pointed directly at her. Everyone laughed, but no one objected. Maybe they thought I would. Why would I start a fight over kissing a boy? Louise and the football star exchanged such a long kiss that some kids got impatient and started kissing the person next to them or got up to get a drink. Whether intended or not, the game was over.

I wasn't sure how I felt about not getting kissed by anyone. I went to my chair and watched as kids began pairing off. Nobody approached me, and since I didn't really know anyone except Louise, I people-watched. Some were kissing; others were talking. I guess they were having fun, but I wasn't. I wondered how it would have felt to kiss the football star in front of everybody.

I was glad when it was time to go. I got my coat and then looked for Louise to thank her for inviting me. She was half lying on the couch with the football star, and they were kissing. I did not think I should interrupt, even though it was rude to go without saying good-bye. I never saw her parents although they may have been in another room. Nobody cared that I was leaving. As I closed the door, I looked back and saw the football star slip his hand inside Louise's blouse. Right there in front of everybody! However, it looked as if no one was paying any attention to them.

That awful night wasn't over. Because there was a car parked in front of my house, the bus pulled up in front of Toto's house to let me off. As I walked back to our driveway, I looked in the car and saw it was the teenager from the first house locked in a boy's embrace. As I went to our back stairway, I wondered why they parked in front of our house.

My mother was in the kitchen crying. "What's wrong?" I asked, in sudden panic.

"How did you get home?" Her voice sounded strange.

"On the bus."

She glared at me. "I saw what you were doing. Go to bed before I say something I'll be sorry for."

I had never seen her so angry, and I didn't know why. I lay awake a long time. That party was a disappointment and not worth Mama's anger. I didn't know it then, but that was the beginning of a lifetime of hating parties, although I went and tried not to show how bored I was.

The next day Mama said, "Who brought you home?"

"I came on the bus."

"I saw you parked in front of the house," she said.

"Oh! That wasn't me. I was on the bus."

"The bus went on past our house without stopping."

I explained about the car and who was in it and that, because it was in the way, the bus stopped at Toto's house. I was never sure if she believed me. The episode was never mentioned again. We both had difficulty putting our feelings into words, and her illness made me reluctant to bother her with my concerns.

I was not curious about how things worked. I didn't need to understand how electricity *worked*; I only needed to know where the switch was. I seldom contradicted anyone even when I was sure they were wrong. It was the same as accusing them of lying. I didn't care what others believed. Only actual proof changed my own opinion.

Until I met Maggie, I never shared my conflicting emotions with anyone. Looking back, I wonder why I didn't confide more in my sister. Was it because she never seemed to make mistakes?

I was anxious to tell Maggie about the party. We had telephones, but they were for adult conversations and emergencies. Since they were party lines shared by four families, you never knew who was listening. I waited until Monday to tell her about the kissing game. She admitted she didn't know how to kiss a boy either, and that made me feel better.

When summer vacation came, I was promoted from kitchen maid to ward maid. Now, because I was 16, I could work more hours, but still on the women's ward. We did not use throw-away plastics. I washed bedpans, wash basins and emesis basins and then autoclaved (sterilized) them.

In those days, beds were lined up in two long facing rows in an open ward divided into cubicles by white curtains hanging on hooks from a railing that ran overhead. When the curtain was drawn, it gave the patient a small measure of privacy. As the nurses got used to me, they allowed me more privileges. I carried trays to a patient, rolled up the head of the bed with a hand crank and arranged the dishes for her.

One day a patient asked me to get her an emesis basin that was out of her reach. As I handed it to her, she started retching. Without thinking, I held her head with one hand and the kidney-shaped basin close to her mouth with my other hand. I had done this for my mother, but it was against the rules for maids to touch a patient. There was no chance to call for help. A senior nurse passed by and smiled. She could have turned me in but didn't. I felt I had gained the respect of these angels in white.

In late August, the nursing director called me to her office. This was like being told to go to the principal's office, a fate I escaped throughout my school years. She began asking questions that I knew had the answers written on a paper she kept glancing at. When I admitted my goal was to become a nurse, she asked, "Do you know what it costs for the three-year training course?"

"No, but I'll work until I have enough."

She smiled. "I have good news for you. Congress passed a law creating a new military service called Cadet Nurse Corps. We need more nurses for our wounded military." I moved to the edge of my chair. "The government will pay your tuition, buy your books and uniforms, and pay your live-in expenses," she said. "It's a wonderful opportunity for girls who don't have money to pay for their own education."

I stared in disbelief. Because of a war where others sacrificed their lives, I was being given a chance I would never have had otherwise! If the government paid for training, I could start as soon as I was old enough.

As if she had read my thoughts, she said, "They lowered the starting age from 18 to 17, so if your parents sign the contract, you can start training when you graduate from high school. There's a class starting next June. They have reduced it from 36 months to 30 months of formal training." I was still trying to digest everything when she added, "The government will give you $15 a month allowance for incidentals during your nine-month probation. The amount goes up after that. On your part, you agree not to disgrace your uniform. When you graduate at the end of 30 months, you must either join some branch of the service or work at a veterans hospital for a minimum of two years."

When I told my parents, Dad recognized it as a great opportunity. He was in favor of anything we wanted to do that led to more education. However, Mama was against it. "You'll be barely 17. That's too young to be on your own."

My golden opportunity was being snatched away. I went to my bedroom and cried until I fell asleep. The next morning, nothing was said, but a few days later Mama said, "We thought about it. We'll sign the contract so you can start in June."

(Later I learned she visited the director and said she thought I was too immature and naïve to be exposed to things nurses did. They came to an agreement. Mama would sign the papers; the director promised I would not be assigned to a men's ward.)

When summer ended, I couldn't wait to share the news with Maggie. I was entering my senior year at sixteen. The first person I met was Louise who asked if I heard about the Cadet Nurse Corps. She would start in June, too. She would be 18 and didn't need her parents' consent.

Maggie wasn't in any of my classes. I looked for her at lunch, but she was not there either. Why wasn't she at school? Mama said I could use the telephone, but Maggie's phone was disconnected. She never returned to school. I never knew what happened. I felt guilty about being so involved with my own life that I didn't try to contact her all summer. That was when I realized there is a *responsibility* attached to friendship.

I concentrated on keeping my grades up. I enjoyed working after school, so my senior year passed quickly. Those of us going into the service (including the Cadet Nurse Corps) took our final exams in May. Training started the first of June although high school graduation was not until mid-June. We would be given time off to attend.

The senior prom was also moved to May. No one invited me. Louise bragged that three boys asked her and she couldn't decide which to accept. "If no one asks you, I'll see if I can get someone to go with you," she offered.

"No. Don't." I didn't want her rejects. I wanted someone to ask me because he *wanted* to!

Mama saved Alice's prom dress. I don't know if she sensed I was hurt or was thinking of the things she missed in her own childhood. She said, "Graduating is an important part of life. You'll regret it later if you don't go. Why don't you ask Buddy?"

I had known Buddy forever. We were on the same youth group bowling team. "He joined the service. He's probably gone," I said. I wasn't sure if I wanted to go with him. I went to the next youth meeting, and he was there in uniform. I asked if he wanted to go to the prom with me and was surprised at how quickly he accepted.

He asked when it was, and when I blurted out the date, he said, "Yes. I can get leave. What do I need to do?" He seemed so pleased that I told myself he would have *asked me* if he had known about it.

Buddy didn't have a car. Going by bus would be awkward and involved walking partway, so Dad drove me to pick up Buddy and then took us to the gymnasium. He would come back and pick us up when the prom ended. This was during gas rationing, so having to drive from our house outside the city limits to Buddy's house and then the school gym meant using extra ration coupons. Dad didn't complain.

Buddy was wearing his uniform and pinned a gardenia on my dress. We spotted a table for two and sat down. Then a strange thing happened. Girls I didn't even know were dragging their dates over to say hello as if we were friends. It didn't take long to realize Buddy's uniform was the attraction.

"Where did you find him?" Louise whispered. She insisted we join her and her date at their table, although her date didn't seem pleased with the arrangement.

Buddy danced with me once. The rest of that long evening I sat at the table watching girls act silly over him. Whenever he sat down, some girl found an excuse to come to our table. It was obvious they hoped he would ask them to dance. Louise didn't wait to be asked. She *asked him* even though girls *never* do that.

They danced several times while her date glared at them. I wondered why her date didn't ask me to dance since we were sitting alone at the table. I guess he was too busy watching Louise. Once he got up, walked across the dance floor, and tapped Buddy on the shoulder. Louise did not look pleased.

That night I learned what makes people popular is not who you are but *who others perceive you to be*. If Buddy had not been in uniform, these girls would have paid no attention to him.

When Dad came to pick us up, Buddy was standing with Louise and two couples. He had his arm around her waist. Her date had vanished. I walked over and told Buddy our ride was here. He waved at Dad and then pulled me aside. "These guys want me to go for coffee with them. I know you have to go with your father. I hope you don't mind. They have a car and will take me home. I'll call you on my next leave."

I walked over to Dad. "Buddy's going with them." He never said a word, but I knew he had sized up the situation and was hurting for me. We rode home in silence. Mama never asked about the prom. I put the gown into the charity box.

I never understood why it was so hard for me to make friends. It seemed as if I didn't fit in with any group. Was it because I wasn't pretty? Looking at my pictures, I was more attractive than I thought at that time. Why didn't anyone ask me for a date? What was I doing wrong? Did my ignorance about the realities of life put up an invisible fence I didn't know how to tear down?

Whatever the reason, I never took part in the high school social life others talk about as being a happy time in their lives. I was glad my childhood was over. I was 17 and about to begin living my dream of becoming a registered nurse.

Kay graduates from Cambridge High and Latin school, June 1944.

CHAPTER **2**

Cutting the Umbilical Cord (1944—1946)

I SLIPPED ON the brown dress that was my student uniform. Using white stud buttons, I attached stiff white cuffs to sleeves that stopped above the elbow. Over the dress I wore a stiff white bib that attached to a wraparound white apron. I already had on brown stockings and shoes. I smiled at the reflection of a girl, now 17, who was starting to live her dream. I hung my summer and winter cadet uniforms in the closet. I was proud of them because they made me part of the military. They also had benefits like paying half price for movies.

We were assigned two girls to a room. As if preordained, Louise was my roommate. She wanted the top bunk, and the bottom one suited me fine. In the room to our left were Becky and Connie; on the right, Doris and Sally. Becky and I became best friends. We were the two youngest students to be trained at this hospital, but I think our real bond was because neither of us had any high school social life. I worked after school; she took care of younger siblings.

Doris was the only married one in our class. Until the Cadet Nurse Corps, most hospitals would not accept married students, believing their minds would not be focused on training, and they could get pregnant. Doris's husband was in the army, so weeks went by without hearing from him. Before the war ended, he was injured, and she left training to care for him.

In spite of our high school friendship, the difference between Louise and me became more noticeable. She was sophisticated; I was naïve. She got better grades, even though I was the one who studied every evening. I felt she was always judging me, and I fell short of her approval.

Our first two weeks were spent in the classroom. That was when my name changed from Catherine to Kay. To help teachers learn their students, we each said our name aloud. I always said, "Catherine with a C, not a K." To tease me, my classmates started calling me "K." That nickname stuck, and I used it the rest of my life. It is the only name most people know me by.

In addition to basic subjects of science, math, biology, and chemistry, we learned how to make a hospital bed with someone in it and give a bed bath by practicing on each other. After the third week, we would work on a ward every morning from seven until noon, making beds and giving bed baths to real patients. After an hour for lunch, we had classes until five and then went to supper. This made a 10-hour workday. Breakfast, supper, and studying were on our own time.

For the first six weeks, we were not allowed to leave the hospital grounds. After that, we were allowed one pass a week (5 p.m. until 10). After the nine-month probation period, we had every other weekend off except when on some assignments and "tours" at other hospitals.

A bell sounded at 9:50 p.m. warning us to go to our rooms with lights out at 10:00. The first bed check was at 10 p.m. and then every two hours until morning during our nine-month probation period. Matron would open the door with her key, flash her light on our bunks, and if all looked normal, quietly close the door. I seldom heard her after her first round.

Both Becky and I were barely 17. She got straight A's in bookwork while I struggled to keep up. I excelled in working with patients; Becky made countless mistakes because her heart wasn't in it. She wanted to be a teacher, but that required a college degree her parents could not afford. When they heard the government would pay for nurse training, they decided that should be her career.

Becky's reputation of being a scatterbrain began when we were learning to give bed baths. After demonstrating, the instructor divided us into two teams. One was assigned to be nurses and would be graded for setup and procedure. Then we reversed roles and the nurses became patients.

Becky thought she was to be a patient first, so she took off her clothes, put on the bed gown, and reported to the assigned bed only to find her teammate already in it. The instructor was so angry she made Becky give the bath and make the bed wearing a gown much too short for her tall frame. It was missing a bottom tie, so there was no way to prevent her backside from being displayed in an embarrassing way.

Unfortunately, it became a hospital joke. Upperclassmen made insensitive remarks to a girl whose self-esteem was already very low. Yet, it wasn't her only mistake. I hoped she would not give up and tried to encourage her.

She encouraged me, too. I struggled to maintain a "B" average in spite of the studying I did. Dropping from the "A" student I had been throughout high school was a shock. Math and chemistry were never my strong points, but my biggest challenge was learning the names of bones and muscles. I couldn't pronounce them, let alone spell them. I spelled them the way my Yankee accent made them sound. It was the hours taking care of *real patients* that I lived for.

One evening Buddy called and asked me to go out with him. I had not heard from him since the prom fiasco. It was five months past my seventeenth birthday, and this was my *first date ever!* (I never considered the prom as a date.) I was excited about being asked.

He didn't object to my getting back before ten o'clock. We took the bus to the bowling alley where we had spent many past hours during high school. Then he suggested walking back to the hospital along the river. That was my favorite walk.

Somewhere along the way, he took my hand. It felt nice. He talked about what he would do when the war was over and listened to my hospital experiences. We were almost at the hospital walkway when he stopped and pulled me in front of him. "I've wanted to kiss you all evening."

Wow! A kiss from a real boy and he wasn't a stranger. But we didn't gaze into each other's eyes first. I guess that is just in the movies. Somehow it changed from locked lips to him forcing his tongue inside my mouth. Panic set in, but he held me tight. I couldn't free myself from his body or his mouth. When he released me, I took off running. I never looked back although I heard him calling my name. I ran up the stairs sobbing and burst into my room.

"My God, what happened?" Louise asked.

"I just had sex!"

I rushed to the sink and rinsed my mouth with water. The words tumbled out like blood gushing from a severed artery. ""He put his tongue inside my mouth, and it went all over the inside, and I couldn't stop him, and I couldn't get loose."

"Get in bed," Louise said. "We'll talk after bed check. Stop crying or Matron will start asking questions."

A few minutes after Matron shut the door, Louise said. "I can't believe how dumb you are. You don't know anything about sex, do you?" she accused. "I'm going to get Doris. I'll get Becky, too. It's time you grow up. Doris will explain it better."

Becky, Doris, and I sat on the floor in front of my bunk. Louise sat cross-legged on my bunk. The darkness was a protective shield.

Louise said, "Kay got French-kissed and thinks she had sex."

Doris turned a flashlight on a book she brought with her. "This is our biology book. Obviously, you haven't looked past our assignment. There's a chapter about sex near the back of the book. You both should read it."

"Does it tell how sex works?" asked Becky. Aha! She didn't know either. That made me feel better.

Doris explained how babies are conceived. I was 17 years old, and the strange thing was that until Buddy put his tongue in my mouth I never thought about *how* you got pregnant. It was something that happened after you were married. Doris answered our questions without embarrassment. I felt like an idiot. But why didn't anyone talk about it? I wondered if my sister knew. Surely, if she did, she would have told me.

It was close to the next bed check when Doris and Becky left. Louise said, "Kay, when you have intercourse, you'll know it." That was reassuring, but how did she know?

A few weeks later, Louise said, "I helped you when you needed it, because that's what roommates do. Now I need your help. I have a date tomorrow and 10:00 is too early to come back."

"Matron will report you for coming in late."

"I'm not going to get a pass or use the front door. I know how to keep the fire exit door open wide enough so it doesn't latch. Nobody uses that stairway to the fire exit door. Matron probably checks it when she does the first rounds. That's why I need help. I need you to put this folded (black) paper between the door and the latch to keep it from fully closing."

"That's not a good idea. If anyone sees you sneaking in, you'll get expelled." I didn't mention it could jeopardize me, too.

"Don't worry. It works. No one will ever know. Do you think I'd do it if it wasn't important? Keep our door locked and don't let anyone in our room tonight. When you go to the fire exit, if you see anyone in the corridor, pretend you're going to the bathroom. Come on. I'll show you. After you fix the paper, go to bed. I'll lock it again when I come in."

Louise had a brown wig and two extra pillows in her closet. I think she wanted the upper bunk because she was less visible there. I watched her arrange the pillows under the covers and place the wig with dark hair showing. From the doorway, it looked like someone sleeping.

I was nervous walking barefoot down the corridor with the paper tucked inside the waist of my pajama bottom. After I placed the paper the way she demonstrated, I hurried back to my room. I did not see anyone.

Minutes turned into an hour, and then it was midnight. After Matron's second bed check, I was like a worried mother. What was taking her so long? Then it was two o'clock check and Louise wasn't back. Had I closed the door too far? What if she couldn't get in? I told myself I'd wait 20 minutes and then check the door. I was still watching the clock when Louise came in. I could breathe again.

"It was easy. Go back to sleep." She fell asleep long before I did.

She had several important dates over the remainder of our probation period, and each time I helped her. I no longer waited up for her, but I was nervous going to the fire exit even though I never saw anyone.

New Year's Eve of 1944 was a somber event. Churches throughout the country held special prayer sessions to end the war. Yet more men died every day. Even a whole nation praying together made no difference.

In January some of us were assigned to pediatrics for six weeks, and nursing was no longer fun. I hated hearing children cry because there was no way to comfort them; they didn't understand why they were hurting or why their parents abandoned them in this hostile place.

That was only part of it. A bigger stress was the pediatric supervisor who was highly regarded by doctors for her competence. Students saw another side of her. Her reputation of punishments was passed on from one class to the next. Being forewarned did not help. It was the uncertainty of what to expect each morning when we reported for duty. We lived in fear of arousing her temper.

My first encounter with her anger was when I cleaned a crib after the child was discharged. I followed the procedure we'd been taught. When the supervisor came to check, she ripped apart my carefully made bed down to the rubber sheet that protected the mattress. "Did you wash that sheet?"

"Yes." I tried not to sound nervous.

Grabbing the back of my hair she pushed my face into the rubber sheet saying, "Do you call that clean?" Before I could answer, she pulled my head up and smashed my face down again. "Does that smell clean? Does it? Does it? Does it?" Each time she said those words she pushed my face further into the rubber sheet. I thought I would smother but dared not resist. Finally she released me. "Do it again and this time do it right."

As I repeated all the steps, I wondered what she smelled that I could not detect.

The second time I got in trouble was while feeding a seven-year-old boy. No food could go back on the tray unless the supervisor approved. This boy ate everything except his chocolate pudding, which he hated. I asked if he had to eat it since it was dessert.

"Is pudding part of the meal?" she asked sarcastically.

"Yes," I said uneasily.

"Then he will eat it."

That was easier said than done. This boy was as determined not to eat the pudding as the supervisor was that he did. I pleaded with him. I coaxed him, knowing she was watching me. Then a doctor came in and she turned her attention to him, buying me a few minutes. I didn't have time to eat the pudding, so pushing my apron aside; I dumped it into my uniform pocket, hoping it wouldn't leak out. When I walked past her to put the tray on the rack, she looked up and started to say something when the doctor asked another question.

I slipped quickly out the door and hurried to my room, hoping my other uniform was back from the laundry. It wasn't. I removed the soiled uniform, dumped the pudding into the toilet, and washed the pocket area the best I could. I hoped my apron would cover the water stains. I skipped lunch and went to the empty classroom. When classes ended, I dropped my books so I could be the last to leave.

The final incident before my pediatric tour ended involved both Becky and me. We worked in the nursery the last two weeks of our pediatric tour, and I loved it. The head nurse explained things without yelling at us. More important, although she was over the nursery too, the pediatric supervisor seldom came there.

We were down to our last few days. Becky and I were scheduled to do our first tube feeding on a premature baby. We were nervous. Doing it to a tiny baby wasn't like practicing on the doll. We heard a rumor that the head nurse was sick. Who would supervise us or would the procedure be postponed?

The pediatric supervisor walked in while we were washing our hands. Oh, no! I turned off the water, grabbed a towel and headed for the glass partition that separated the preemie nursery.

She had decided Becky was an inferior student. Becky's best would never be good enough, and today was no exception. Even behind the glass partition, I heard her voice rise accusingly, "Are you using *hot* water?"

I didn't hear what Becky said, but whatever she did or did not say enraged the supervisor. She turned off the cold water, leaving the hot running. I watched with cowardly horror as she grabbed Becky's lower arms and held her hands under now scalding water.

Becky was crying. I should do something, but I was too scared. The supervisor yelled, "Get out of here and don't come back!" Becky fled, her hands dripping water.

In seconds, the supervisor was approaching me with cruelty written on her face. Her eyes were wide with a rage that made me tremble. She followed me to the crib. I picked up the feeding tube. I don't know if she saw me trembling or if I did something wrong. She snatched the tube out of my hand and pushed me with such force I fell to the floor and slid across it on my butt until I landed up against a wall.

"Get out of my nursery!" she screamed.

Scrambling to my feet, I ran out the door. I found Becky sitting on the floor in the corridor tunnel to the nurses' quarters. She was looking at bright red hands and sobbing. I was crying, too.

"You have to go to the emergency room. Your hands look burned."

Becky said between sobs, "They'll ask what happened and they won't believe me. Everybody thinks she's God. It must be my fault." Then she stood up. "Go to lunch. I'm going to my room."

"You need someone to look at your hands," I said to her retreating back.

She didn't answer. I watched her turn the corner, feeling a surge of guilt. I knew I should offer to go to the emergency room with her to back up her story. I was as scared as Becky. The pediatric supervisor was respected by doctors and everyone in high positions. She saved her abusive side for students afraid to complain.

I tried to eat, but a knot in my stomach wouldn't let me. I went to Becky's room. The door was locked, and she didn't answer my knocking. I pleaded with her to open the door. When there was still no answer, I hoped she had gone to the emergency room. I went there and checked behind each curtain. No one asked why I was there. Classes would start soon; maybe she was there.

Becky never came to class. The instructor asked if anyone knew why. I spoke up. "She wasn't feeling well. I think she's getting checked by a doctor."

When classes were over, I knocked on her door again. When there was still no answer, I went to the dining room and saw Connie getting in line. "Have you seen Becky?"

"Not since this morning."

I told her what happened. "Do you think she's asleep and didn't hear me?"

We went together to their room, and Connie unlocked the door. Becky was not there. Her clothes in the closet and dresser drawers looked untouched. Her clean student uniform and her two cadet uniforms were there, so she still had on the same work uniform. We never left the hospital wearing our work uniform.

Becky did not return to her room. According to Connie, Matron never questioned her absence at bed check. In the morning, while Connie was getting dressed, a maid came and asked which drawers and closet were Becky's. While Connie was on her assignment, someone packed Becky's belongings in boxes. They left her student and cadet uniforms hanging in her closet. If Becky quit, why didn't she pack her own things? Neither instructor asked about her.

After classes, I went to their room with Connie. The packed boxes were gone and so were her uniforms. The next day, Connie asked the head nurse on her ward, and I asked an instructor about her. We were both told to mind our own business. Sally checked the newspaper, including the obituaries. There was no mention of an accident. None of us ever heard from her again. Why didn't she leave me a note? My guilt for not taking her to the emergency room haunted me for years.

This was worse than losing Maggie. Another best friend disappeared without a trace, and I didn't have any to spare. I hoped she went home, told her parents, and they notified the director so the pediatric supervisor would have to explain what she did. I couldn't understand why Becky never wrote to me.

We finished our nine months' probation on my 18th birthday. The hospital held a dance in our honor. Students from Harvard University as well as servicemen in Boston were invited. Bob, an attractive sailor, asked me to dance. I felt an instant attraction. We danced or sat and talked the rest of the evening. He expected to be shipped out in two days, but he asked if I'd go out with him tomorrow. Except for Buddy, it was the first time anyone asked me for a date.

Bob was waiting in the nurses' lounge. We took the bus to a small café for supper. He asked if I wanted to go to a movie. I wanted to spend the time getting to know more about him, so I suggested we walk back along the river. We talked about past experiences and future dreams. I liked him and could tell he liked me, too.

He seemed to be as naïve and bashful as I was. It was getting close to the warning bell when we reached the nurses dormitory. We stood in the shadows near the door, holding hands. I know he wanted to kiss me, and I wanted him to. Every time we drew close, another nurse came hurrying by. We jumped back like kids caught stealing cookies.

The warning bell rang while we were waiting for the right moment. Reluctantly, I said goodnight and went inside. I thought of and worried about Bob often over the remaining months of war. We wrote letters that connected us even though we were from different backgrounds. I prayed God to send his ship to Boston again. To my knowledge, it never came back.

During this period, my mother's cancer was recurring. Every time she thought she won a battle, the cancer showed up somewhere else. Now that we had every other weekend off, I went home. On Sunday, Dad drove me back to the hospital, and we sat in the car talking. I felt closer than ever to him. We talked about education. He felt he had been cheated out of a better life because of lack of education and didn't want it to happen to his children.

He said when he immigrated to the States that he hoped to travel to all the states. Falling in love, having a family to support, and a sick wife in a research program pushed the dream into the attic of his mind where it remained. I shared his wanderlust and planned to travel after I graduated. I was lucky. A nurse's license is a ticket to go anywhere; there are always jobs for nurses. "You kids will have to do my traveling for me," he said. I promised I would.

Like Dad, my mother did not have a happy childhood. In her teens she played the mandolin and enjoyed performing. That stopped when she married, had three children close together, and then was sick the rest of her life. Her dream was to own her own house, but even with a better job, Dad could never save enough. Then he heard about a bungalow in a small town that Mama would love. Banks were more generous with loans, but first he had to have the down payment.

When Mama's uncle died, Aunt Dell inherited a large estate. Since they had no children, his will said his wife had the use of all the money until she died. What was left would be divided between his former medical school, Mama, and another niece. I thought if I explained it right, she would let Mama *borrow* enough from her eventual share of the estate for a down payment.

When I called and suggested a visit, Aunt Dell said, "You are welcome to visit, but I have no room for overnight guests. Come at noon and I will take you to a nice restaurant for lunch."

I bought a train ticket to Maine. Not wanting to spend money on lodging, I took an early train and, after lunch and a visit, planned to take the train back the same afternoon.

As soon as I arrived, I had a feeling my visit was a mistake. She was having one of her "bad days," so her niece prepared lunch in the apartment. Watching the dynamics between my aunt and Mama's cousin, I realized Aunt Dell didn't appreciate that her niece was spending her life taking care of someone who treated her like a troublesome servant.

Lunch consisted of tripe, which I will never eat again, fresh tomato slices, white bread, butter, blackberry jam, and tea. That's it. I forced myself to swallow the tripe, trying not to gag. When the painful meal ended, I began my plea.

"Mama's cancer is back. We don't think she has much longer to live. Dad found the cottage Mama dreamed of. Please let her borrow some of the money Uncle Howard left her in his will. Dad can make the monthly payments. He only needs the down payment." Aunt Dell did not interrupt, so her reply stunned me.

"When people are *given* something, they don't appreciate it as much as if they work for it." (She *earned* her wealth by marrying a doctor.) "Your father will find a way if it was meant to be."

"If it was meant to be." I hate that excuse for transferring to God your unwillingness to help. I started to answer when her niece shook her head warning against it.

Aunt Dell said, "We're finished. Clear the table and wash the dishes thoroughly this time." When she turned to me, I saw hate in her eyes. "Obviously, you came to get my money. You have your facts wrong, like most children. Your mother doesn't get one penny until I die. If she's as sick as you claim and dies before me, her share goes to charity. Your father won't get anything, and he knows it. That's why he's sent a child with a sob story."

I stared at her. It wasn't only her refusal; I could feel hatred filling the room. Did she hate Dad and both nieces?

She interrupted my thoughts. "It's time for my nap. You can see yourself out."

When I told Dad about the visit, he said, "If I'd known, I'd have told you not to go. She never forgave your mother for marrying an uneducated foreigner with no family she could brag about. Don't worry about us. Your graduating is what is important."

(Several years later, Aunt Dell died at home with her niece still caring for her. I hoped the niece got her share. If Aunt Dell could have changed the will, I bet she'd have had the money buried with her.)

The world began changing rapidly after we hung the 1945 calendar. On April 12, Franklin Delano Roosevelt died of a cerebral hemorrhage. Sixteen days later, Mussolini was shot "while trying to escape." Two days later, on April 30, Hitler and his mistress/wife (Eva Braun) committed suicide inside his bunker. Two days after Hitler's suicide, Berlin fell to the Allied Forces, but VE Day wasn't proclaimed until May 8, bringing an end to the war in Europe.

I wrote Bob telling him how happy we all were to see that war end. But for Bob, whose ship was in the Pacific, the war was not over. More lives were being lost on both sides. One of them could be his.

On June first, I finished my first year of nurse's training. Now we spent more time working with patients than in the classroom. I thought how this would have affected Becky and realized she would eventually have quit. I wondered, as I often did, what path her life had taken.

In August, our class was divided again into groups that took turns working six-week tours in obstetrics, surgery, and the emergency room. Except surgery, which was day duty, some girls worked evening and night shifts. We still worked 10-hour days, including classes, but we had every other weekend off. Studying was still on our own time.

On August 6, after several warnings for Japan to surrender, our country dropped the first atom bomb on Hiroshima, obliterating half the city. Japan still did not surrender. Three days later, the U.S. dropped its second atomic bomb on Nagasaki. There were rumors that Japan would surrender, but it seemed nothing changed. A week passed.

On August 15, I took the subway to Boston to go to a movie. I wore my cadet uniform so I could get in for half price. As I reached the top of the subway steps, a fire engine came by loaded with sailors and civilians. Horns were honking and people were cheering.

"The war is over! The Japs surrendered!"

Someone yelled, "Climb aboard!" A sailor reached down. "Let me help you." Impulsively, I took his hand and people made room for me to sit. "My name is Jake." He was attractive, with reddish blond hair and a nice smile.

We rode up and down the streets, singing, cheering, and shouting until the fire crew had to get back to their station. They let us off in front of a bar, so several sailors and girls headed inside to continue celebrating. Jake pulled me along with them. Someone pushed the tables together, and we all sat down.

I had never been in a bar before and was only 18, but no one was checking IDs. Jake asked what I wanted. To cover my ignorance, I said, "Whatever you're drinking."

I didn't like the taste or the way it made my head feel. I felt warmth inside, which was partly the drink but mostly being accepted by a group of strangers who seemed like old friends. I had only half finished my drink when someone called for a second round. I stood up. "Sorry, guys. I'm a student nurse and have to get back to the hospital before 10 o'clock."

They tried to persuade me to stay. "It's a special occasion!"

I knew Matron wouldn't see it that way. Jake walked me to the door. I suddenly realized he was not much taller than I.

"We'll be shipping out. Will you write to me?" After we exchanged addresses, I hurried to the subway.

The mood of the country became jubilant, but hospital life continued on schedule. Both Louise and I were doing our tour in surgery. She was better than I was, which seemed to be the story of my life. I was too short to see what the surgeons were doing, so I could not anticipate what they needed next. When surgeons brought their own nurses, we provided a stool if it was needed. Student nurses did not have that privilege.

Because the next surgery couldn't be set up until the room was emptied of all evidence of the prior one, everyone helped gather linen and push tables with dirty instruments into the utility room. While housekeeping washed down the walls and mopped the floor, the nurses headed for the break room for coffee and that much-needed cigarette.

Since I didn't smoke and drank very little coffee, I started washing instruments and preparing them for sterilizing. The others knew I was doing their work. I expected them to thank me, but no one did. Since I was the only one on our crew who didn't smoke, maybe they thought I should do the work.

When I complained to Louise, she was not sympathetic. "It's your own fault. Smoke with us and we'll come back earlier and help."

"I don't know *how* to smoke!"

"It's easy. Just put the cigarette in your mouth and inhale," she demonstrated. "Here, try one of mine."

When I tried to inhale, I went into a fit of coughing.

"You'll get the hang of it with practice," she encouraged. "Don't be a fool. Take a break with the rest of us."

Cigarettes were sold everywhere and cheap to buy. Nobody knew smoking caused lung cancer, and no one suggested secondhand smoke affected anyone. In fact, smoking was encouraged by advertisements and movie stars; cigarettes were included with military C rations. You could smoke almost anywhere.

The next day I went with the others and lit a cigarette. The head nurse stared at me. I couldn't enjoy sitting there because we had work waiting, but I took Louise's advice and waited with the others. The head nurse jumped up. "We've stayed too long. We've got work to do." Was it my imagination that she looked at me?

I went to the break room all that week. Sitting in the smoke-filled room, I felt uncomfortable and the conversation was inane. On Friday night, I told Louise, "I hate sitting there when I know we should be getting the next surgery ready."

Louise shrugged. "Do as you please."

I felt chastised. That ended my smoking career. I gave Louise the rest of my cigarettes. Until the end of our surgery tour, I did the majority of cleanup between surgeries. No one seemed to appreciate it, but I was glad I never got addicted to smoking.

Christmas of 1945 was celebrated with enthusiasm. Servicemen were drifting home. We rang in 1946 with rekindled hopes and wild celebrations. For me, the best news was Dad felt sure he would have enough money to put down on a house by summer, and the cottage he had found was still up for sale.

I had turned 19 when I was assigned to Boston City Hospital for one of my affiliated hospital tours. Jake, the sailor I met on the fire truck, was back in Boston for extensive ship repairs. We had exchanged letters, but I never expected to see him again. We arranged to meet at the pond at Boston Gardens. I wore my cadet uniform so he'd recognize me.

We talked about what had been happening to each of us. We exchanged kisses, but I didn't feel the sparks romance novels talked about. We arranged to meet again the next week at Boston Gardens. When Jake arrived, he had a nice guitar with him. We didn't have as much to talk about, so he played the guitar and sang songs I wasn't familiar with.

When it was time to leave, he said, "Will you keep my guitar for me? It'll be weeks before they finish repairs. When a ship is in port, things have a way of disappearing. Please keep it for me."

I did not want to be responsible for it, but I could think of no good reason to deny his request. We began dating almost weekly, but I did not enjoy the dates. I asked him repeatedly to take his guitar back, but he always had a reason why he couldn't. In June I started my last six months of formal study, and we were still dating. When I said I had to study, he pleaded and once had actually shed tears until I agreed to another date.

Maybe he really liked me, but nothing about him attracted me. I had continued seeing him because I felt guilty rejecting someone who had risked his life while I had free government-paid training. Every night I prayed he would ship out with his guitar, which had taken on the aspects of an albatross. In mid-June, his ship was still here. Would it ever leave?

I decided I would take the guitar with me on the next date and leave it sitting on the ground if necessary. Before I could carry out my plan, Jake showed up at the nurses' dorm. When Matron told me I had a visitor, I was excited as I hurried to the lobby, hoping Bob's ship was in Boston. Disappointment enveloped me when I saw Jake. I had told him to never come unless he called first.

Disappointment turned to anger. "Why are you here?"

"We're shipping out tomorrow. I have to talk to you. Can we go to your room?"

"No!" I exploded. "We can't have guests in our rooms!"

"Where can we go? I have to talk to you in private."

We left the dormitory and walked down to the river. There was no one around.

"I want to marry you," he said, pulling me to him.

The smell of liquor on his breath was nauseating when he kissed me. His grip tightened and it became a French kiss. Knowing that didn't help me get free. Suddenly I was on my back, his body pinning me down as he pushed up my skirt and tugged down my panties.

I guess in today's world it would be called "date rape." There was no such thing then. Since I went with him willingly, society viewed it as consensual. If you make a bad decision, you live with the consequences.

When Jake sat up, he was saying something, but I was too furious to listen. I couldn't even cry. Jumping to my feet, I readjusted my clothes. "I hate you. I never want to see or hear from you again!"

He was asking for forgiveness, but I turned my back and walked to the hospital. His pleas dogged at my heels. I never looked back. When I got to my room the first thing I saw was his guitar. How would I get it back to him? I was *not* going to see him again. I crammed it under my bed where it would not be a constant reminder of my first sexual experience.

In July, I started getting nauseated. My periods were erratic, and I no longer kept track of the dates. Could I have missed one? Surely, in one incident I couldn't be pregnant. Yet, the signs were there.

After I told Louise what happened, she shook her head in disgust. "Why weren't you using condoms?"

Before the war, birth control of any type was illegal in Massachusetts. Although it still was illegal, because Boston was a main navy port, drugstores sold condoms under the counter. If Louise assumed I had enough control of the situation to make Jake wear a condom, she didn't believe it was forced sex. If Louise, who knew me so well, didn't believe it, no one would. I went with him willingly. That made it my fault.

"If you *are* pregnant, you can't hide it. You should quit before people find out and it gets on your record. Marry him. You can get a divorce later, but it will save your reputation."

I felt like a dog that was kicked and then offered a bone. "Please don't tell anyone," I pleaded.

"I won't, but how long can you hide it?" She looked at me. "I hope you aren't stupid enough to get an abortion from some back alley quack."

Abortion was illegal in Massachusetts and probably in most of the states. During my surgery tour, I watched doctors clean out the mess from a botched abortion by someone without a license who may not have used sterile instruments. Some girls used a wire coat hanger to abort themselves. I knew that after self-abortions, girls could bleed to death, and some could never have children afterwards.

When we discussed it in class, I smugly thought only someone who was ignorant would have an abortion. Now I understood. To hide the shame of being unmarried and pregnant, you might do anything, including risking your life. There is an old saying that goes something like, "Never judge a man until you have walked a mile in his shoes." The realization came that even that couldn't make you understand; chances are, his shoes would never fit your feet.

I think Louise kept her promise not to tell anyone. We had been roommates for two years. This was pay back for the chances I took to help her stay out late. Or maybe it just wasn't important to her.

During July, I prayed every day for a miscarriage. Daily morning sickness was exhausting. I had started my home health tour, which included lots of bus rides and walking as I checked on discharged patients. I taught them to give insulin, helped with treatments like colonic irrigations, and changed colostomy bags and dressings. Sometimes I could hardly keep from gagging when I was doing treatments that never bothered me before.

Normally, home health was a miserable assignment in the middle of summer, but it allowed me to be away from classmates who could observe my daily vomiting. Sometimes I stopped at every telephone post with mostly dry heaves because I was skipping breakfast. I hoped the nausea and vomiting would be over by the time my home health tour ended, but it was worse.

Emergency room was next. I couldn't keep running to the bathroom there. There was no way to continue hiding morning sickness and future body changes. I was pregnant and nothing was going to change that. I had thrown away the dream that guided me since I was a child. The saddest thought was I lost it because of someone I didn't even like.

How could I tell my parents what a failure I was? My graduation had become as important to them as to me. I felt lost and scared and hopeless. "You threw it away!" That self-accusation kept running through my mind.

CHAPTER 3

Consequences (1946 & 1947)

THEY SENSED SOMETHING was wrong when I walked into the living room where they were having Sunday morning coffee. "You don't look well. Are you all right?" Mama asked.

"Something terrible has happened." My voice was shaky from trying not to cry. I looked at the carpet, suddenly aware of how old and faded it was. I blurted out, "I'm pregnant."

A long silence was broken by a floodgate of questions. Who was it? When did it happen? Where did I meet him? Is he in the service? How old is he? Where does he live? I answered the questions that poured from them. Then I said, "I tried to stop dating him, but he said he loved me and I didn't want to hurt him."

Relief settled on Mama's face. "You must marry him."

"You don't understand. I don't love him. I don't even *like* him. I felt sorry for him, but I don't even feel that now."

Dad said, "Does he know you are carrying his child?"

"No. He was gone when I found out." I was crying now. "Louise said if I marry him then I can get a divorce. Is that what I *have to* do?"

"No!" Mama said vehemently. "No divorce!"

"What can I do?" I felt utterly hopeless.

"Put the baby up for adoption," Dad said quietly.

"Yes!" Mama agreed, latching onto this hope. "The church will find a place for you to stay, and they'll arrange the adoption. I'll call tomorrow."

She meant a place to hide in shame. Although I hadn't thought of adoption, it made sense.

"You must tell the father," Dad said. "He has a right to know. Maybe his family will want the baby. Do you have a way to reach him?"

"I have his mail address. I think his ship still comes in and out of Boston."

"Write and tell him to come here. I don't want you to meet him by yourself after what he's done."

Relief washed over me. Dad believed me! I had begun to think no one would.

Mama interrupted. "Does anyone at the hospital know about this?"

"My roommate. She promised not to tell anyone."

"Tomorrow, you must talk to the director. Tell her I'm sick. It isn't a lie. I haven't said anything, but the cancer is back, and I don't think I'll get better this time."

I was shocked. She was always positive about beating cancer. "You can't mean that. You always get better."

"This is different." Her voice was flat, with no sign of the emotion she must feel. "Tell them you are too distracted to study." She sighed. "Maybe someday you'll finish your training."

That was my first ray of hope. If no one knew about the pregnancy, maybe I could still be a nurse someday. That hope carried me through the months ahead.

I wrote to Jake:

"I told my parents about you, and they want to meet you. I have something to tell you and don't want to do it in a letter or over the phone. And you need to take back your guitar, because I'll have no place to store it when my folks move. I'm through training, so I'll be here for the next three weeks."

I added the directions and signed my name. I was trying to find a way around the truth without lying. The next day, I posted the letter on my way to the hospital to pack my things and meet with the nursing director. I managed not to cry, but there were tears in my voice. "My mother's cancer is back. She doesn't think she will get well. I need to leave. I'm sorry."

She studied me. "You don't look well. I wondered what was wrong. Do you realize you only have 18 weeks of formal training? The government is doing away with the program. In just 18 weeks you can ask for a variance to take the exam for your license. "

Even if the nausea stopped, which it showed no signs of doing, I would start showing before that. If anyone suspected, I'd be kicked out. I had another physical examination coming soon. "I wish I could, but I have to go now."

She sighed. "It's a shame to quit when you are so close. When your mother's need is over, come back and we'll see if there is a way to finish your last 18 weeks."

"Thank you," I mumbled, looking down to conceal my tears. If she knew I was pregnant, her sympathy would evaporate. I was beginning to live a lie that would last most of my life.

She sighed. "Leave all your uniforms in your closet. They belong to the government."

I rushed to my room, glad everyone was on their assignments. Burying my face in my pillow, I sobbed. Was it realistic to think I would ever finish a dream so close and yet so far away?

Several days later, Jake called. The conversation was strained, but if he noticed, he didn't say anything. His ship would be leaving again but he wasn't sure when, so I asked him to come that evening. When he arrived, he was the perfect gentleman, reminding me of the night we met.

Before he could mistake this for a friendly visit, I gave him his guitar. Surprise showed on his face. Before he had time to speak, I said, "There is something you should know. I'm pregnant." I was amazed at how easily the words came.

Jake's face turned red. My parents were listening. It surely unnerved him. He bristled. "How do I know it's my baby?"

I couldn't blame him for being skeptical. What are the odds I'd get pregnant the first and only time I had intercourse? But it was the wrong question as far as Dad was concerned. He leaped to his feet, uttering a profanity I'd never heard him use.

Jake, sitting by the door, jumped up, dropping the guitar as he yanked opened the door. Dad picked up the guitar, held it like a club, and swore he'd kill him.

They both ran down all three flights of stairs. Mama and I looked out the window and saw Jake running across the street with a pretty fair lead. Dad was behind him with the guitar held like a club. I think Jake would have run all the way to Boston if need be. Dad stopped, flung the guitar in the general direction of Jake's back, and it crashed onto the pavement. My albatross met its death, and Jake lost his prize guitar.

While Dad sat on the ground to catch his breath, Jake disappeared. When Dad got home, he was still breathing hard. Mama hovered anxiously, "Dick, are you all right? Are you sure you're all right?"

Worried and heartsick at the trouble I had caused, I snuck to my bedroom and cried myself to sleep.

I thought often about Jake in the following days. I even made excuses for him—he thought he was in love with me—he had too much to drink. Yet I could not accept marrying someone I didn't even *like*. We weren't ready for marriage. We never talked about important things like our religious beliefs, but I knew it caused problems in marriages. I knew he wanted to live near his family, and I could not picture myself spending my life with him for the "sake of our child." I doubt he knew how much being a nurse meant to me because we talked only about the things he wanted.

Mama was right about losing her long battle with cancer. Standing all day ironing exhausted her. I packed the things she wanted to take to the cottage Dad had finally been able to get. Moving there was all she talked about.

Mama went with me to our minister, who knew of a church-sponsored home for unwed mothers in another state where there was less chance of anyone tracing the adoption. In those days, birth mothers and adoptive parents knew nothing about each other. This was to protect adoptive parents from attempts by the birth mother or child to get in touch with each other. The birth mother had to care for her baby for one month to give her one last chance to change her mind. It was a fair system. Adoptive parents would be the only parents some children ever knew about.

"The home for unwed mothers" would not take girls before their fifth month and many girls waited till their seventh month. With my nursing background, our minister knew of a place where I could live and work until then. Although "the home" was supported by charity donations, each girl provided her own personal items. Donations provided formula, baby clothes, cloth diapers, and the use of a crib while you were there. No throw-away diapers in those days—you scrubbed out the stains using strong soap and a washboard.

If a mother kept her baby, she could not take any donated items when she left. I wouldn't have that expense, but without government help, I needed money to pay for the remainder of my training if I was allowed to return. Returning was the hope that I hung on to. Being a nurse was still my goal.

I bought two maternity outfits for 25 cents each and an old suitcase for a dime at a charity store. I took two favorite skirts and sweaters, underwear, pajamas, toilet articles, and my biology book. Mama would keep my other books. My government-issued topcoat was gone; but I still had a coat, scarf, wool hat, mittens, and snow boots from before training.

I went first to the home for unwed mothers. Miss Hudson, the matron, would tell me about the live-in job. I hoped she would be as nonjudgmental as our minister.

With only one small suitcase and an over-the-shoulder purse, I maneuvered easily from train station to bus, and the driver promised to tell me when we reached my stop. From there I walked to a large house. It looked ordinary until I noticed the first story windows and front door had decorative iron grillwork reminiscent of prison bars.

When the matron answered the bell, she was tall, with wide shoulders and a square, almost masculine face whose prominent features were a hawk-like nose and a tight mouth that didn't look as if it knew how to smile. Her name badge advertised who she was.

She eyed me critically. "So, you're the nurse. There's no excuse for a *nurse* getting pregnant. I'll get you registered." I followed her through an unoccupied sitting room and into an office. She handed me some papers. "Fill these out." She pointed to a chair in front of a table.

They wanted information on whom to notify in an emergency, pregnancy questions, medical history, and a list of rules I signed my agreement to follow. The last page was an adoption release.

When I gave them back, she checked for signatures in the proper places. "You have made the sensible choice. Adopting parents are screened; they are well educated, established in careers, and good church-going people. They provide a better home than any unwed mother." She studied the paperwork. "You can come at Thanksgiving or wait until January. You must be under supervised care the last two months."

She opened a drawer, and took out an address book, found the page she wanted, scribbled an address on a piece of paper, and handed it to me. "Go back to the bus stop. Wait for a bus with 4 on the sign. It takes you back to the train station. Take bus 21 there and get off at the Home for Disabled Children. I'll tell them you're on the way."

I thought she would show me through this building, but it was obvious she wanted me to leave. While waiting for Bus 4, I thought it was ironic that the service I most disliked—taking care of children—was where I would work for the next four months.

It was after dark when I arrived. The nurse who answered the door was expecting me. She led me to the attic, showed me the bathroom and a dormitory with seven beds. Four girls were sitting on their beds talking and reading. She introduced them and then pointed, "That's your bed in the corner. You start work at six-thirty and work until two, with a half-hour for lunch. You are off from two to four while the kids nap and then work until seven-thirty. We all eat meals with the children."

Most of the girls were my age or younger. One of the older girls said, "You make the sixth aide. Martha over there is a kitchen helper. Jenny went out with her boyfriend, and Vera's on her day off. Everybody has a different day off, but never Sunday because that's when visitors come."

Daisy was the only black girl. She said, "Thought I'd hate workin' split-shift but it don't matter no-ways 'cause there ain't nothin' to do 'round here."

Kris said. "That's a lie, Daisy." She turned to me. "You can take a bus to the city on your day off. Mondays we have liver. I hate it! There's a diner about a mile west. I walk there at Monday afternoon break. They got good hamburgers."

I was glad the girls were friendly. I learned there had been unwed mothers working in the past, but I was the only one now.

Miss Riley assigned me to the babies and younger children, explaining, "Since you've worked in hospitals, you're used to people dying. Daisy goes crazy when a baby dies. The girls forget everyone here is waiting to die."

Most of those under my care had Down syndrome or hydrocephalus. Congenital hydrocephalic babies have fluid around the brain that causes an enlargement of the head. At that time it was commonly called "water on the brain," and there was no treatment. Except for mild cases, babies didn't live long. Today shunts and other techniques allow a hydrocephalic baby to live a normal lifespan.

Although Daniel had a severe case of congenital hydrocephalus, he defied the odds and was still alive at 18 months. By now, his head was so large his little body could not support it. He lay on his back day and night. His parents had dropped him off when he was two weeks old. They paid his board but never came to see him. Death instructions were on his chart and, in big letters, "Don't call family: *notify by mail*." Cremation was prearranged with an undertaker.

Daniel had a lovable grin and, strangely, did more smiling than crying in spite of what must be constant pain. He said a few nonwords. Ma-ma-ma was what he called everyone, pee-pee-pee when he needed his diaper changed, and din-din-din when he was hungry. That, plus crying, smiling, and pointing, were his means of communication.

I couldn't cuddle him like other babies, so I massaged his legs, arms, and lower back and talked to him whenever I had time. I started with, "I'll tell you a story about Daniel." It was the same story about two beautiful angels with snow white wings who would take him in their arms, and when they did, his head would stop hurting. I showed him pictures of a blue sky with white clouds where he would go to live. When he wanted me to tell him that story, he said "stow-ree-ree-ree."

Babies with Down syndrome were my favorites. They are such loving children who want to be hugged and are always ready to smile. Life is strange. I have a grandson with Down syndrome. He is proud of the medals he won in Special Olympics swimming competitions. He has grown into a sensitive, loving man-child who remains a joy to his family.

I had been working seven weeks when I dreamed Mama was calling me, but when I tried to reach her, a river blocked the way and there was no bridge. I ran along the river bank trying to see her on the other side, but there was only her voice that became increasingly fainter. I awoke in a cold sweat.

When I dozed off again, the dream came back, but this time her voice was coming from our grandparents' house. When I opened the door, it turned into a tunnel like nurses use to get from the dormitory to the hospital wards. I kept running but the tunnel never ended. This time when I woke up, my heart was beating like a sledgehammer. I could not go back to sleep. I packed my suitcase and sat on the side of the bed waiting for morning.

When I found Miss Riley, she looked at my suitcase. "Is something wrong?"

"Yes. It's my mother. She called last night. Her cancer is worse. She needs me." In my mind it was true.

"I thought you didn't want people at home to know you are pregnant?"

"We moved. No one in the new neighborhood knows us. Anyway, I have to go. I'm sorry to leave without notice. Maybe I'll be back."

"It's starting to rain. You shouldn't stand waiting for the bus. The maintenance man can drive you to the train station. Your job will be waiting if you return. Everyone thinks you're a good nurse." I thought it was nice of her to say that.

When I arrived in Boston, I called Dad at work and told him where I was. "That's strange," he said. "I was going to call you and Alice tonight to see if one of you could come home. Your mother is very sick. I don't think she can last much longer. I'll pick you up as soon as I can leave here."

On the way home, we talked about Mama, who was in her dream cottage but unable to do any of the things she wanted to do. Six months earlier would have made a big difference. She was on codeine, but it didn't help much. When we got home, she was standing at the stove, stirring a stew for supper. Her stomach was far bigger than mine.

She looked up. "You're here. I hoped you'd come. I think I'll lie down." She never got up on her own again.

Dad moved some of his things to make room for mine in their room. He would sleep on the couch. He said, "This doctor knows she doesn't want to be in a hospital. That's why I was hoping you or Alice could come. I hate leaving her alone, but I have to work and Roy can't do it."

After observing how bad the pain was, I called the doctor. Doctors, especially in small towns, made house calls to patients too sick to get to their office. After checking her over, he thought she'd be more comfortable if he drained fluid from her abdomen. Her bed was too low to work on, so they carried her to the table where she sat on pillows with Dad and Roy supporting her back. The doctor withdrew over a pint of fluid. I asked if she could have something stronger than codeine. "I'll give her morphine before I leave."

"I've had two years R.N. training. I can give injections."

"Good," he said. "I'll leave the syringe and this bottle of morphine." He turned to Dad. "I'll leave a prescription for more morphine. Get it filled on your way home from work." As he handed me the bottle, he squeezed my hand, "You know, don't you?"

I wasn't sure if he meant she was dying or to give her as much morphine as she wanted. With the strict code of ethics doctors and nurses work under, I didn't ask questions. Whatever he meant, I was determined she would not suffer as she had been.

The next night, with morphine to deaden the pain, Roy and Dad carried her to the living room. Mama told us about the curtains she planned for the windows. "I wanted to make them myself, but there are some just as good in the Sears catalogue. It's in the second drawer of my nightstand."

A page was turned down. The four of us looked over the pages, and she pointed out the ones she wanted. Her face brightened. The curtains seemed to be a milestone. "I'm tired," she said. Dad carried her to bed and sat with her until she fell asleep.

"I'll go by Sears after work tomorrow," Roy offered. "If they don't have them in stock, I'll have them special ordered." He was measuring the windows.

"No," I said. "If they don't have them, get something close. Special orders take too long."

The next night, Dad carried her to the couch so she could supervise hanging the curtains. These weren't the same as those in the catalog, but if she noticed, she never mentioned it. Roy had also bought curtains for the kitchen and bedroom. Because there were no curtains up, I realized she had been much sicker than any of us thought when they moved here.

After they carried her to bed that night, she never left her bed again. She slept a good deal the three weeks I was with her. With no one dictating when she could have pain medicine, I gave it to her whenever she asked. Perhaps because she knew she could have it when she asked, she paced herself, so there were times when we were able to talk. During those snatches of time, we made up for years of poor communication and came to understand each other. I was glad I came home.

She ate very little—a few bites of scrambled eggs, custard, or junket. Mostly she wanted hot tea, but never more than a few sips at a time. I called the doctor to ask if we should give her intravenous fluids.

"Don't you think it's time to let nature take its course? We medical people have interfered with God's plan long enough. Give her ice chips when she's awake, and put Vaseline on her lips to keep them from cracking. Give her food or fluids when she asks, but don't force them on her."

Morphine made her feel strange, so she thought I was giving her liquor frozen in the ice chips. She kept telling me not to let Dad know she was drinking. There was no reasoning with her. She *knew* she felt different. I started showing her the liquor bottle which was always at the same level. She thought I added water so Dad wouldn't know, and that satisfied her. The only obvious change during the last week was that she needed morphine every two hours and was still in pain but was sleeping more.

After supper on Sunday, she asked Dad to take me to the evening church service. Before we could refuse, she said, "I want to talk with Roy."

"I'll give you another shot first." I reluctantly stood up.

"No. It fuzzes up my mind."

This was the first time I had been out of the house since coming home. I didn't want to go. Maybe she hoped Dad would go to the service with me, but he sat outside in the car. When we got home, Roy met us at the door.

"She's gone," he said. "Twenty minutes ago."

I couldn't believe it. How could she die while I was gone? Didn't she know how much it meant for me to hold her hand at the end? "She could have held on another 20 minutes *if she wanted to!*" I said angrily. I was mad at myself for leaving her.

As I walked into the bedroom, three things struck me. She was holding rosary beads between her fingers, and there was a look of peace on her face. The third was a crucifix nailed to the wall over her head. I had never seen it before. She had given up her Catholic religion when she married my father because of my grandmother who was brought up Anglican (similar to Episcopal). I suddenly realized her reason for wanting to talk with Roy was that he had become a Catholic. We didn't discuss his decision, but we all accepted it.

"What happened?" I asked, sitting on the edge of the bed so I could put my hand over hers that were already cold.

"She wanted me to pray with her," Roy answered. He looked at Dad. "She wanted me to tell her it was okay to become a Catholic again. I gave her my rosary beads. Did you know she had that cross hidden in a drawer? We were saying the rosary together when she stopped, gave a big sigh, and was gone. It was so quick."

He turned to me. "She felt you were hanging onto her, and she was tired of fighting. That's why she wanted both of you gone."

An autopsy was required by the research program. The report revealed every organ and bone in her body was consumed with cancer. They published her treatments and death in a medical journal. The article ended by stating radiation allowed her to live far beyond anyone's prediction. For those who loved her, her life was too short. I wonder if, living in a body riddled with cancer, it seemed to her that life was too long.

Alice was at work when Dad called, but quickly got someone to replace her and took the next train home. There was no funeral except that a priest met us at the funeral parlor and conducted a brief service for the immediate family.

Dad insisted on cremation even though her choice would have been burial. Even Roy agreed it should be his decision. Dad took the urn of ashes by himself and scattered them somewhere that had a special meaning to him. Alice left for college; I went back to New York, leaving Dad and Roy to mourn together.

When I walked into the nursery, the first thing I saw was Daniel's empty crib. I went to our dormitory. Lying on my bed, I cried for a little boy nobody wanted. Was anyone holding Daniel's hand or saying a rosary for him? Did he see the beautiful angels I said would come for him? I wish I could have been there, telling him his stow-ree-ree-ree when he died. And then I remembered all the suffering he had endured in his short life. He had finally escaped.

My own baby kicked me hard, reminding me there was another unwanted child. Tears became sobs that continued until my head ached. I was crying for Mama, Daniel, my baby, and my own lost dream.

Later, I asked when Daniel died. "A few days after you left," an aide said. "He kept saying sorry-ree-ree-ree like he was apologizing for dying."

Miss Riley overheard us, "His passing is a blessing."

While I was gone, the relief nurse quit. Miss Riley needed a replacement and offered me the job until she found someone permanent. As it turned out, I held that job, which included two day shifts and two night shifts, until I left in January. The added responsibilities came with a raise even though I was only working four days, 12 hours each, instead of 10-hour days six times a week. The difference in pay for aides and qualified nurses was impressive.

From my first week there, I had been going to the diner on Wednesdays because it was kidney stew day (worse than tripe). With more days off, I asked if they could use part-time help. The manager said he couldn't afford a waitress. "Can you afford free coffee?" I asked. "I work cheap."

I was not an official employee, but I got free coffee and a hamburger whenever I went there. When I waited on customers, I could keep any tips, although people seldom tipped in diners. More important than the money, I got experience serving orders, keeping coffee cups full, and running a cash register. Someday I might need it. Even more important, I learned there are always jobs to be found if your expectations are reasonable. It helped pass the time, and I enjoyed conversing with adults.

On cold, rainy days, I never went to the diner. Now the manager offered me two dollars a day to work on my days off. After that, neither bad weather nor my increasing size kept me from walking to the diner.

The holidays were a blur. Some children had no visitors on Christmas, so we tried to make it a fun day. The older ones must have wondered why they had one present from Santa and others had several. If they asked, what would I say? Because their family loves them, and we're the only ones who love you because that's our job? I'm glad no one ever asked.

In January, I packed my few belongings. Miss Riley had the maintenance man drive me all the way to the house for unwed mothers because it was a bitter, cold day. She was always thoughtful, and these kids under her care truly needed that.

When I returned to the unwed mother's home, Miss Hudson was more welcoming, although she still did not smile. Her world was black and white, and "the girls" were sinners. What was there to smile about? She did a necessary job for her church and did it efficiently. She was never unkind, but it was obvious she had no respect for those who sin—something I doubt she did in her entire life.

She showed me through the house and explained the rules again: no visitors; get permission to walk to town; always be at the table on time; clean up after yourself; no snacking between meals; eat everything on your plate; and lights out at nine o'clock. Televisions were still in the future, but I doubt she would have let us watch one.

There was a large upright radio in the sitting room we could listen to on special occasions; we had to listen to the church service Sunday morning. That was its purpose. We were given a Bible to read Sunday afternoon. There was also a small collection of books, mostly stories of missionaries, left by former clients. I never saw anyone reading them. Romance and Western novels were passed secretly between the girls. If discovered, they were confiscated as trash. I kept my biology book hidden in fear she would decide it was pornography.

A doctor came every week to examine girls who had a problem, but his nurse saw everyone, recording our weight, vital sounds, frequency of bowel movements, and any complaints we might have. Whatever doctor was on call at the hospital would deliver us when our time came.

The dormitory for those who were pregnant contained 12 beds with a small footlocker for personal belongings. One clothes closet was shared by all occupants. It contained maternity outfits left by people who had passed through the system; if you borrowed a dress, you washed and ironed it and hung it in the closet. We could put our own maternity dresses in this common closet or keep them in our locker. Most of us hung them in the common closet. I thought it was a good system.

There were two dormitory rooms on the second floor: a small one for mothers keeping their babies and a larger one for those up for adoption. Miss Hudson explained the downstairs beds were taken, but as soon as one of the girls went into labor I would get a bed. Until then, I would sleep in the room upstairs with mothers whose babies would be adopted.

Beside each mother's bed was a small hospital-type bassinet for the baby. Whether you kept your baby or gave it up, you had to care for it for one month before you could leave. I spent as little time there as possible. Hearing those mothers crying made me wonder if I really could give up this baby that was now constantly reminding me of its presence.

Barbara's bed was next to mine. She was 32—the oldest girl there—and was giving up her son because the father was a married senator. She had been his executive secretary and was in love with him. She thought he would leave his wife and children for her, but when she got pregnant, he blamed her and insisted she get an abortion that he would arrange and pay for.

When she refused because of her religious belief, he said they were through. She suggested putting the baby up for adoption, hoping that would pacify him. He insisted she come to "the home" at five months; she did whatever he wanted in a futile attempt not to lose him. He sent a generous check and a formal work reference. Before she went into labor, she learned he had a new mistress. He never intended to leave his wife and children.

After the baby was born, in a last desperate attempt, she used the birth of a son as an excuse to call him. He refused to accept her call. Forced to admit it was over, she knew this might be the only child she would have. She decided to keep him but had not told Miss Hudson. She began nursing him while still in the hospital.

A tremendous emotional battle was raging between my heart and mind. Adopting parents were considered saviors for unmarried girls. Instead of welfare, there was scorn and an ugly name for the child if it became known the mother was never married. Adoption made sense, but emotionally it was tearing me up. I would never know what kind of people they were. Educated, of course, with stable careers, a house, and church-going Christians, but did that make them loving parents? If Miss Hudson was married, she would qualify as an adopting parent. I didn't want my baby to grow up as narrow-minded as she was. How could I go through life never knowing what was happening to my own child?

I realized I could never finish training with a baby to support. I could work as a nurse's aide, but I now knew the pay was half what an R.N. made. I could probably make more money on tips as a waitress. Where was the emotional satisfaction in being a waitress? I didn't know what to do. I called Dad and told him about my emotional struggle.

"I thought finishing training meant everything to you?"

"It did, but I didn't expect to have a baby. I don't think I can give my baby away. What should I do?"

"I can't make that decision for you. But whatever you decide, I'll help as much as I can. If you keep the baby, come home. We'll figure out what to do."

That was all I needed to hear. I think he knew I'd regret it all my life if I went through with an adoption. My bed was downstairs now, but I headed upstairs to tell Barbara about my decision. I stopped on the stairs when I heard her arguing with Miss Hudson, who apparently caught her nursing the baby. I heard Barbara say she was keeping her son.

"Oh, no, you're not!" Miss Hudson yelled. "I already have the adoptive parents. They've signed the paperwork. This baby belongs to them!"

"Let's call my attorney. You said this month was for me to be certain I want to give up my baby. I'm keeping him. I signed temporary papers, but the *father* did not. He has friends in Washington."

Miss Hudson must have realized this could have a nasty conclusion. Fathers were not involved with the girls who stayed in homes for unwed mothers, but did they have legal rights? Would adoptive parents want this baby if it meant a court fight?

"Take your bastard and leave," Miss Hudson snapped. "I want you out of here tomorrow."

When Barbara left, she said, "Don't worry. I have friends and enough money." I gave her Dad's address, but she never contacted me.

I had signed the release for adoption. I needed to tell Miss Hudson now before she found adoptive parents. This was not the way I planned my life, but I realized keeping my baby was something I had to do. When I told Miss Hudson, she was nicer than I expected. Maybe it was because no adopting parents were involved, or maybe Barbara's remarks worried her. However, she was not encouraging.

"This will complicate your life more than you realize. What will you tell people? You need to pray on whether you are making the right decision." She took my release out of my file and put it under a slot called PENDING. "Remember, decent men don't want to raise an illegitimate child. Adoption by good families is *always* best."

Some girls went home with their babies, but most walked away with nothing but a suitcase. All of us felt we made the best choice in a situation where there was no *good* choice. I am glad society now accepts single mothers, but for me it meant I would be living a lie.

Then it was my birthday. When I turned 20, they had a birthday cake for me. Everyone sang the traditional song. I made a wish and blew out the candles. I appreciated that Miss Hudson tried to make it a special day, but it wasn't.

I went into labor two weeks later. They called a taxi and sent me off with my file of paperwork. I don't know why I was scared. I had worked in both labor and delivery and knew what to expect. The truth is, when it is *you* having that baby, it isn't the same as watching.

This was a busy obstetrics service and appeared understaffed. I was in a labor room by myself and I began to wonder if they had forgotten me. I had no one in the waiting room asking about my progress. The pains were getting stronger and closer. Unable to stand it any longer, I rang for a nurse. When one answered my bell I said, "I thought you had forgotten me."

"We don't forget our patients," she said, with a smile meant to reassure me. I knew it could happen. She checked me and said, "You have quite a way to go yet." She gave me a pill and I fell asleep. I came fully awake at the sound of someone screaming and realized it was me.

People in surgery gowns began buzzing around. They wheeled me into the delivery room. Someone told me to move onto a table. Through the pain, I heard voices yelling, "Push, damn it, PUSH!"

I was suddenly consumed with a pain so severe it tensed all my muscles and wouldn't end. It felt as if I was being torn apart. Then, just as suddenly, my muscles relaxed and I took in a deep breath. A nurse held up a squalling baby girl for me to see while the doctor repaired my torn pelvis with an episiotomy. I had seen it happen to other first-time mothers.

There are no words to express the feeling of holding my own tiny little girl in my arms. Why did I ever think I could give this baby away?

The time in the hospital and at the home passed quickly because Sherry kept me busy. When it was time to go home, Miss Hudson walked us to the door where I would begin the next phase of my life. She cautioned, "If people learn you have an illegitimate child, they'll shun you. Make up a good story. And keep your legs crossed when you go on dates."

There was no story to make up. I would say Sherry's father was a sailor and never came back. Proof of legitimacy never came up. Times were changing. There were too many babies whose fathers never came home from the war. Many single women were raising children, and people no longer asked if you were a widow or a sinner. Yet Miss Hudson was right about one thing. With the loss of so many men, those who would make good husbands and fathers were at a premium and preferred women who didn't have someone else's children.

Now, almost three years after starting my nurse training, I had a little girl instead of a diploma. I didn't know how I would support her, but I believed with certainty that I would find a way.

Would I have been so sure if I had known what the future held?

CHAPTER **4**

Becoming Resilient (1947 & 1948)

LIFE IS FULL of choices. We make what seems the best choice because we cannot predict the future. I believed keeping Sherry would mean ending my dream to be a registered nurse. I could still work as a nurse aide which was my favorite part of nursing. I would just have to work more hours because the pay was so low.

When I arrived home, I had a letter from Grandma Mel, who had just turned 78. She offered to stay with Dad and take care of my baby while I finished training! I never dreamed of such a generous gift and especially from one whose life had been so hard. She gave me a second chance to make my dream come true. How lucky I was to have a family who loved me in spite of my sin. I called Grandma to thank her and make transportation arrangements.

I didn't know if our community hospital would take me back, but I knew Boston City would. Growing up poor taught me not to waste money on things I didn't need. Now I'd be even more careful. We found a crib at a charity resale store, as well as baby clothes and a dozen cloth diapers, which we would reuse. As for myself, I needed nothing except work uniforms, and I had to wait and see which hospital accepted me. Each hospital had its own color and style uniform for students.

After Grandma got settled, I rode with Dad to the subway. The director seemed glad to see me. "I'm sorry your mother died. You'll be glad you returned. Finding a boyfriend or a quick way to earn money traps girls in a career they never wanted."

"Then you will take me back?" I asked, eagerly.

"Yes. The Cadet Nurse program is over, but since most of your training was paid for, I think the board will waive the last few months training fees. I put your duty uniforms in storage hoping you would return. You don't seem to have gained weight, so they should fit. Do you still have all your books?"

"Yes, ma'am."

She studied a file. "You have finished all your specialty tours except emergency room and the six weeks at Worcester State Mental Hospital. The rest will be either the medical or surgical wards. I think we should get your emergency room tour over first. When did you want to start?"

"As soon as possible."

She laughed. "How about next Monday? That should give us enough time to set you up."

I was so happy that I didn't mind going to the emergency room, even though I always worried about doing that tour where life-or-death decisions must be made quickly.

It appeared Louise had kept my secret. It was evident the director had not heard. I decided to remain as unnoticed as possible. It didn't work out that way.

I should have had a chest x-ray before starting work, but when I went to get it, the x-ray department was overrun with victims of a car crash involving kids on an outing. The technician asked if I'd come back on another day, and I willingly agreed.

I got involved in emergency room life-and-death situations and forgot about the x-ray. I was in constant fear of making a serious mistake.

It was two weeks before the x-ray technician found my uncompleted chart and tracked me down. It was just a routine chest x-ray we had every six months. My mind wandered as I waited for the technician to tell me it was okay to get dressed. I was beginning to think he had forgotten me when he returned, "The film was blurry and they want me to take it again."

Was it really taking longer than normal or was I just impatient? Finally, the technician said I could get dressed, but I could not leave until the doctor saw me. Was this some kind of new rule they had? In a few minutes, the doctor called me to his office.

He began by asking me a lot of strange questions. Then he said, "There is a suspicious spot on your right lung. It isn't well-defined and you say you don't have any other symptoms, so I'll let you continue working. We'll take another x-ray in a month."

It had to be the blurry film because I felt fine. The next day, my confidence eroded when I was called back to the x-ray department where two doctors waited.

"There is a suspicious spot on your right lung. We don't think it is anything to be alarmed about, but the emergency room may be too stressful while we monitor it. You are being assigned to night duty on the women's medical ward for a month. During that time, you will go to the emergency room to have your temperature taken every morning and at four o'clock in the afternoon. You are to remain at the hospital on your off-duty time. Is that understood?"

I was glad to get out of the emergency room, even if it was temporary. Night shift is an easier assignment, but I found it difficult sleeping days. I hated not being able to go home and hold Sherry, but I called every day.

I felt fine—a little tired perhaps from working again, but I wasn't worried about x-rays. I wasn't sick. I looked forward to the restriction ending so I could see Sherry.

When it was time for the next x-ray, as soon as I finished my night shift, I went to have it taken. After taking several views of my chest, the technician told me to wait until he checked them out. No sense getting dressed if the doctors wanted more views. They should get their equipment fixed, I thought. It seemed to take forever for the technician to come back, and I wanted to go to bed.

A doctor came in and sat on the x-ray table facing my chair. "You definitely have tuberculosis, although it is in an early stage. If it stays confined in your lower lobe, you have a good chance of recovery. Good luck, young lady."

Before I could even digest his words, a student nurse arrived with a wheelchair to admit me to the same ward I had been working on. I was put in a private room used for isolation. Infection control gowns, masks, gloves, and basins of disinfectants were already in place. I must be dreaming. Everything was happening too fast.

I was falling asleep when the resident doctor arrived and began asking family health questions. Then he called a student to assist while he did a physical exam. When finished, he dismissed the student. After she left, he pulled a chair beside the bed and sat down. "How old is your baby?"

I was stunned. I stared at him. How did he know?

"The episiotomy scar is new, so it wasn't long ago. You know it's my duty to put this on your chart? If I do, everyone in the hospital will soon know. You're a student with a highly contagious disease. Other students will be curious and look at your chart. No way to prevent that. All it takes is one person to notice the episiotomy notation. You know how fast hospital gossip travels. Does anyone know you've had a baby?"

I shook my head, still too shocked to speak.

"Do you want to tell me about it?"

I told him my story, where Sherry was, and how I've wanted to be a nurse all my life. It felt good to tell someone who really listened.

"You may never be well enough to finish training unless they find a cure. It is a shame because we need nurses. Most hospitals won't take a chance on anyone who has had TB, and you know trainees with small children aren't accepted anywhere. That's two strikes against you. Your medical and training records will follow you for a long time. I'm sticking my neck way out if I don't record it, but I'll make you a deal. If you promise to keep your mouth shut, I will too."

Tears were streaming down my face as I thanked him. I was so upset with having my secret discovered and remembering his promise to not record it that I forgot to ask about treatment.

Someone brought a lunch tray. I couldn't eat. I hadn't slept since yesterday and I was tired. Yet I couldn't sleep. My mind was jumping from one thought to another. This had to be a bad dream. Then two students dressed in gown, headgear, and mask helped me put on similar gear before they wheeled me to my dormitory room to get a few personal things.

"You can't take clothes except your pajamas, robe, and slippers. Someone will pack your things and send them to your family."

"I'm not sick!" I insisted. "I'll be back when they discover their mistake. There's no sense sending my clothes home."

"We do what we're told. You know that. Please hurry. I need to get off on time today, and this is putting me behind."

I couldn't think what to take, so I grabbed some toiletries, a notebook, and my biology book. I was back in bed and finally dozing off when a social worker said, "Ready to go for a ride?"

I got dressed again in the gown, headgear, operating room slippers, and a mask to cover my lower face. The things I had taken from my dormitory room were inside a laundry bag. She wheeled me to a white car with "Social Services" in black writing on the side. She tried to make polite conversation about something of no importance until we came to a large building called Tuberculosis Sanitarium. I was transferred to a wheelchair and taken to a private room.

A nurse came in with the familiar isolation equipment. She handed me a clean hospital gown and took away what I'd been wearing. "I'll put your robe and slippers in the closet, but you won't need them or your pajamas for a while. You're on strict bed rest. That means *stay in bed.* Ring your bell and we'll bring you a bedpan."

I was getting scared. They were treating me as if I really was sick. When she closed the door, I felt as if I was alone with my world on the other side. I was wide awake. I looked at my watch. It was eight hours since I had reported to the x-ray department that morning. I had not even had a chance to call Dad and tell him what was happening. I vaguely remember someone saying he would be notified.

I felt the same as always. I was *not* sick! How could they make this terrible mistake? How long would it take them to realize I didn't belong here? How long would Grandma stay? I should have gone through with the adoption. What was going to happen to Sherry now? What was going to happen to me? Questions I couldn't answer were pushed aside as I tried to remember what we had learned about contagious diseases.

Tuberculosis, or consumption as it was often called, is a bacterial infection that was considered incurable and was highly contagious. In the old days, people died of it at home. Now, if they discovered you had TB, you were separated from the general population and sent to a sort of modern leper colony called a tubercularium. Tuberculosis (TB) can be spread by a cough, a sneeze, exchange of sputum through kissing, or eating with contaminated forks and spoons. It was as much feared as leprosy. Of the two, TB was more communicable.

(Tuberculariums were built in cities or counties everywhere in the '30s and '40s. Although streptomycin was discovered in 1942 by Seiman Waksman, it was not introduced until 1946. Antibiotics used in treating tuberculosis as well as other diseases were not developed from streptomycin until after 1952. When tuberculariums were no longer needed, they were turned into small community hospitals, county homes for the disabled, or nursing homes for the elderly. Until the 1960s, "active" tuberculosis was considered incurable and carried a death sentence. That is still as true in many countries around the world as it was centuries ago.)

TB struck children as well as adults in their prime. As a student, I had never seen a patient diagnosed with it because, like me, they were immediately rushed to the nearest tubercularium. Many people were ashamed to admit they had a relative with it, and, like polio, everyone was afraid of catching it.

The treatment in 1947 was bed rest, pneumothorax (pneumo) to collapse a diseased lung, or surgery to remove a lobe if the disease was limited to only one part of a lung. In addition to lung tuberculosis, there was a deadly form of bone tuberculosis, which was seen in rural areas where folks still drank unpasteurized milk from their own cows. *(All the patients at Cambridge Sanitarium had lung tuberculosis.)*

The only way I could accept my sudden isolation was by believing it was an x-ray equipment failure. They would find out when they did further testing. If I really had TB, my chances were considered non-existent for getting well. It had to be a mistake. How quickly would they discover their mistake? Grandma was 78. How long could she take care of a baby?

My thoughts were interrupted by a knock, followed by the door bursting open. Three people stood there grinning at me. "Hi, we're the welcoming committee."

Because of the fear of catching TB, many people wouldn't visit family members in a tubercularium. The patients themselves formed small support groups that we called gangs. Everyone faced the same fears when they were admitted, so those patients who were "getting better" formed groups of three to six people who visited a new patient on the first day. Each gang became close friends who encouraged each other. These three people would become my gang, and we would hang out together whenever possible.

They were three very different-looking people. The first one I noticed was a six-foot, good-looking redheaded guy in his mid-twenties. He had lots of freckles. He introduced himself as "Red," and I felt an immediate connection. I must be attracted to redheads.

The second was a girl about my age with long blonde hair, blue eyes, and an angelic face. Her name was Pat.

The third was difficult to describe. Peggy was tall and very skinny. She looked like a scarecrow that had lost its stuffing. A patch of hair on top of her head looked like a haystack. I thought she was an old woman, but I later learned she was only 35. Her two front teeth were missing and the rest stained yellow by long-time tobacco usage, even though she no longer smoked. The most noticeable thing about her was her ugly smile, and she smiled constantly. Does she realize how silly she looks? I would never smile if I looked like that. I was still trying to figure out this strange old woman when Pat asked, "What have they told you?"

"They have more tests to do. I don't have TB. It was bad x-ray equipment. How long have you been here?"

"I'm starting my third year."

Wow! I felt like someone had just thrown a bucket of ice water on me. I would learn patients were brutally honest with themselves and each other.

Red said, "I was here a couple of years my first time. This is my second relapse. I get pneumo every week now, so I may be here a long time."

Peggy read the distress on my face. *"Don't compare yourself with other people*. We're all different. Hey, I'm the only prostitute here—'less you count Dolly. She was a stripper and gived it away." She doubled up with laughter. What was so funny about strip dancing or prostitution? I would have been ashamed to admit it.

Life is strange. Peggy—a prostitute—became my best friend. In an unexplainable way, she took the place of my mother. I would come to love all my new friends, but Peggy most of all.

My denial about TB didn't last long. I never felt sick, so it was hard to imagine my lungs were infected, but x-rays don't lie. They took another x-ray and a sputum test. My new doctor gave me the devastating news. As soon as he left, Peggy, who must have been waiting outside my door, rushed in. "What'd the doctor say?" She was not smiling.

"He said I'd be here *six months* or a year. Maybe longer. Six months is a *long* time."

I didn't tell her how worried I was about Sherry. My grandmother had agreed to stay for five months and two were already gone. It looked as if I'd never finish training. Most important, I should have gone through with Sherry's adoption. It would be even harder to do now. Most adopting parents prefer to adopt newborns. Would Grandma stay until I got home in six months?

"Tell me the truth, Peggy. How much longer do you expect to be here?"

She shrugged. "Till I die." It was said without emotion. She added quickly, "Hey, don't feel bad for me. I got it easy. Three squares and a warm bed. Save your pity for the girls on the street tryin' to make a livin'."

I began to sob. She held my hand and waited till I was cried out. "Doc didn't mention pneumo, did he?" I shook my head. "Well, see, that means you ain't serious sick. You can be outta here in no time."

"You don't understand," I blurted out. "I have a three-month-old baby at home. I'm all she has. Her father was a sailor and never came back." She nodded as if she believed me. "I could have had her adopted by people who would give her a good home."

"Why didn't ya?" Peggy asked.

It hit me like a ton of bricks. "Because I *wanted* to be her mom. I didn't want strangers raising her, maybe abusing her. It was my worst mistake yet. What will become of her now?"

"I'm gonna tell you what I tell all the new girls workin' the street. 'Ya gotta take it *one* trick at a time.' That's what you gotta do, Kay. Take it one day at a time."

I never forgot that advice. I was learning to set priorities, and my first was getting well. I was glad I told Peggy about Sherry. It was good to confide in someone. I never asked her to keep it secret, but I knew she would. She was the kind of person who thought if I wanted people to know, I would tell them myself. She never gossiped and didn't have a mean bone in her body. I came to love this brutally honest woman with a genuine smile that no longer looked ugly.

Living in a tubercularium must be similar to living in prison. The outside world disappears. All focus is on the daily activity of the place you live and on those who live there with you.

Dad visited twice a month and told me Grandma would stay as long as she could. I assured him it would only be a little longer because the doctor said I'd get out in six months. Dad started to say something and then changed his mind. After a short silence, he said. "She'll stay as long as she can, but Sherry is getting more active. If she has to go before you get well, we'll figure something out. You let me worry about that. Concentrate on getting well."

Dad was my only visitor. My sister was living in Thailand; my brother couldn't come because of his precarious heart condition. I thought how similar his disease and mine were. We both *felt* perfectly well, we both *looked* perfectly well, and yet each of us lived with a silent killer in our chests.

Roy sent me rosary beads with instructions. It made sense to cover all options, so I said the rosary faithfully every morning and night and prayed for a quick recovery. I forced myself to eat everything on my tray (even the liver served with eggs for breakfast twice a week). I never cheated by getting up when no one was looking, and I forced myself to stay positive. I concentrated on one thing—getting well quickly.

With our system, after you had three negative sputum tests plus two stable x-rays, you were given specific time up. All my sputum tests were negative and there was no dramatic change in my x-ray, so after two months of complete bed rest, I was allowed to get up.

For two weeks, it was one hour a day, and then two hours. A month later, I was up two hours in the morning *and* two in the afternoon. Because of my quick progress, I convinced myself the TB diagnosis was a mistake. By the start of my sixth month, I was up four hours in the morning and evening in preparation for going home. I never stayed up longer than I was allowed, although it would have been easy to do.

Radio was all we had, but each patient had earphones with the ability to select one of three stations. The tubercularium also had an extensive library, so

I read many books.

I used my "time up" to visit the gift shop where toiletries, candy, and stationery were sold, to attend the Wednesday night movie, visit bed patients, or eat a meal in the dining room instead of on a bed tray. Most important, I could use the only telephone and call home. This allowed me to keep daily contact on Sherry's progress.

In the same week, my gang added Tony and a girl named Winnie. We split our time visiting each other, gathering at the bedside of someone on bed rest to play cards. I got to be a whiz at cribbage as well as whist. I also learned to knit and crochet.

Peggy was my best friend from the beginning. Her comments were never judgmental. When I confided the sailor never married me, she said, "What's done's done. *Ya can't let the past control your future.*" That advice stuck with me through the bad years ahead.

She told me how she came to be a prostitute and always tried to give good service for money paid. It paid better than any other job she could get with her lack of education. Peggy loved showing newcomers the ropes and helped those who needed a temporary place to stay. Her life philosophy didn't change at the tubercularium. She was the first to find the right words of encouragement for new patients.

When I asked about her missing front teeth, she shrugged. "Sometimes ya get a customer mad at life who feels like he gotta hit somethin' and your face is there. Most guys are happy when they leave. Sex is a business like buyin' anythin' else."

Peggy smiled all the time, even when she'd been dealt a bad hand, whether it was cribbage, which she played with passion, or being moved back into a private room when she started coughing up blood. She had not told the doctor about her chest pain or fatigue, symptoms of advancement of her disease. She knew—we all knew—she'd never leave that room.

While Peggy was back in isolation, I was advancing toward recovery. I visited her for an hour every morning and evening, wearing a gown, mask, and surgeon's gloves. We talked a lot and played cribbage. She usually won. I no longer thought of her as a prostitute.

Many patients formed attachments with the other sex. Pat liked Tony; I had a crush on Red. This involved writing notes and holding hands on movie night. It helped to know a guy I liked actually liked me, too.

I had been in the tubercularium nearly six months and was eating all my meals except breakfast in the dining room. On the next doctor's visit, he said

I could go home. My disease was in remission. Not "cured." You were never considered cured.

Before I left, I went to see Peggy, knowing I'd never see her again. The tubercularium was more than 75 miles from where we lived, and it was not on a bus line. I wasn't supposed to, but I leaned over her bed and hugged her. Her toothless smile lit up her face, and I marveled at how beautiful it was.

"Kay, didja know you beat the odds? You're outta here in six months! I ain't never heard of nobody doin' that. Ain't never heard of nobody gettin' out without surgery or pneumo. Do ya know how you done it?"

"By following the rules."

"That helps, but most of us do that. When Doc said you'd be here *six months, a year, maybe longer,* he didn't really *mean six* months. He was just tryin' to take ya down easy. All you heard was it can be done in six months, and by God, you done it. You honest to God beat the odds! I'm proud you're my friend." She grinned. "Fact is, if I could, I'd send all my best customers to you."

Peggy died shortly after I left. I wrote to her regularly, but she never answered. Maybe she couldn't. I wondered if she ever went to school. My last letter came back unopened with "deceased" sprawled across the front. Winnie died next followed shortly afterward by Red. Pat lived a couple of years longer. Of our gang, I was the only one that I thought beat the odds. I didn't know it, but TB wasn't through with me yet.

I learned so much about life and death at the tubercularium. Not just from Peggy, although she was my mentor. That uneducated, penniless prostitute taught me more about being happy than all the successful people I've known. She taught me that what people look like and the work they do is not important. People are different, so don't try to compare them, and never compare yourself to anyone else.

I learned if you truly believe you can do something, it is possible to beat the odds against your success. You simply learn how to fight it "one trick at a time."

I learned *not to worry about how much time I have left, but make each day important.* I would think of that often in the future when my life spun out of control.

When I left tubercularium life behind, it was Christmas again. So much had happened in 1947. I went to the home for unwed mothers, had my first baby, went back in training for six weeks, and spent the rest of the year in a tubercularium.

In January, Grandma left. This kind lady lived 22 more years—the last 10 of which she was blind. She was still mentally alert when she died at 101.

While I was away, Roy introduced Dad to Gladys, the bookkeeper who worked in the same office as Roy. They had become close. Gladys had a girl about five years of age who had been boarded with a family almost since birth. She recommended we put Sherry there while I finished training. The doctor gave me a letter recommending I rest for two or three months on limited activity. If my x-ray was still stable in three months, I could go back into training.

Roy was having heart problems, and his doctor recommended he take a couple of months off work. After a family conference, it was decided Roy and I would stay with our aunt and uncle for a couple of months. So Roy, Sherry, and I went to New Hampshire. By now, my family wanted me to graduate as much as I did. The time passed swiftly in New Hampshire.

The last weekend before we were scheduled to come home, there was an important hockey game between the local New Hampshire team and the Boston team where so many players were Roy's personal friends. My aunt, uncle, and Roy went to the game. They said it was a nail-biter game with first one team scoring and then the other. There was screaming of encouragement and insults from fans on both sides. Roy was totally caught up in it. At the end of the game, he had his first heart attack.

We came home the next weekend as planned, but Roy was an invalid after that and was in and out of our local hospital several times. *(Almost exactly two years after my mother died, he died in the hospital the morning he was scheduled to go home again.)*

After we got home from New Hampshire, I went back to the hospital to see if they would let me finish training. "You are determined, aren't you?" the director said when I handed her the doctor's recommendation. She insisted I spend the first six weeks working a half day which lengthened my training but was probably a wise decision. Sherry went to live with the people who took care of Tina. Dad paid her board while I finished training.

In June, I went to Worcester State Mental Hospital for my last two months of training. It was a traumatic period that gave me nightmares for years afterwards. I started on the women's ward. The uncomprehending look in their eyes made me nervous, but I wasn't afraid until my third day of duty.

It was "shower day" for the ward I was on. This meant staff guided or forced reluctant patients down a long hallway to the shower room. There the reluctant ones were stripped of the canvas dresses that covered their nudity. No one wore underwear.

Attendants wearing bathing suits took over. Some patients went under the running water willingly, seeming to enjoy it. Others stood there shivering until an attendant came with a bar of soap and a long handled brush and pushed them under the water where they scrubbed them down as if washing a reluctant dog.

One student passed each one a towel to dry with and sent them to my station. My job was to give each one a clean dress. Sound simple? There were mostly pink dresses, but every so often there would be a blue one left from a previous time.

One woman decided she wanted the blue dress she saw halfway down the pile. I wasn't going to give it to her, but she glared at me and screamed, "Give me that dress!" It didn't seem worth fighting about, so I reached down into the pile and gave it to her.

Wrong decision. Another woman decided it *was* worth fighting about and tried to take it from the first lady. Then everything happened so fast that I don't know how I ended up on my back on the wet cement floor looking up at a dozen wet, naked women above me who were screaming, pulling each other's wet hair, or exchanging blows. Most had no idea what the fight was about.

There are no words to describe the nightmarish feeling of being trampled by countless *naked* flying feet, legs, bodies, and arms. Then a naked body fell on top of me. When I tried to push her off, other naked arms with hands attached pulled at my hair and uniform. Now I was fighting too.

Help arrived. Somehow the whole mess got sorted out, and I was left wet and humiliated. It seemed such a simple task. How could I have got it so screwed up? The supervisor called me aside and asked what started the fight. After I explained, she said. "*You* must always stay in control. Get cleaned up. You can take the rest of the day off. You look like you need it."

I wondered what I should have done to stay in control. On the way to my dormitory, I had to go through several locked doors. In my uniform pocket under my white apron was a skeleton key that allowed me to unlock and relock each door. Some inmates thought it was funny to pack the key hole with feces. Sometimes they packed feces in so tightly, you had to dig it out little by little with your key. I had forgotten to take tissues in my pocket to clean the key. When I found two blocked keyholes and heard people laughing at me, I broke down crying.

I dreaded going to the ward the next day, but when I arrived, everyone had forgotten the incident except me. The supervisor sent me into the locked room to monitor patients' actions and make notes on their charts as to what they did during this social time. I was afraid of all inmates now, and being behind locked doors with them terrified me. They could be calm and then, without warning, turn violent.

Some stared at me with blank eyes and others with hatred. I tried to stay out of the way of those who walked back and forth. A few sat on the floor, swaying in invisible rocking chairs; others stood by the small barred windows looking out. What did they see that held their attention? They were nameless, so I had no idea what to write on the charts. Gradually, during the next three weeks, I became less afraid. I could identify some from passing out medications. Now, I could make real notes and not just reword the previous notation.

One day a fight broke out. I pushed the panic button, and the staff came running. They worked in prearranged teams, each team grabbing one of the fighting women. Someone yelled at me to grab a leg and help carry one of the fighters who was screaming profanities.

To calm them down, each woman was stripped and then lifted onto a wet sheet covering a stretcher. We then wrapped them as tight as a mummy in the wet sheet so they could not move. The woman I was wrapping continued screaming profanities and spitting on us if we came within range. She was helpless against trained attendants. There were actually three women who would lie there for hours while they calmed down. As the sheets dried, they became tighter. I thought it was a cruel punishment and asked one attendant why we did it.

"Because it works," she said. "After a few times, most learn not to fight."

The other treatment I participated in that made me feel physically ill was electric shock. By the way patients fought when we were restraining them, I knew they realized what was going to happen would be hellish. We used a combination of leather cuffs at ankles and wrists to restrain them to the stretcher, a belt around their waist, and a leather strap around their head to keep it under control and prevent head injuries while they were convulsing. They tried their best to fight against that strap around their head, which meant to me they knew what was about to happen.

Once the head strap was in place, someone pushed a padded tongue depressor between the patient's teeth, and said, "Ready!" We stood back while the electricity was turned on for only a couple of seconds. "Hold!" I grabbed and held rigid the assigned arm or leg while the patient convulsed violently for what seemed like an eternity. It was my responsibility to prevent the leg or arm bone from breaking during the convulsion.

I only had to participate in that a couple of times. I'm not sure how often this shock therapy was done, but the same doctor came to shock the designated patients. They determined those needing a shock treatment by the handwritten notes nurses made on their charts and the patient's responses at group therapy sessions. I understand that electric shock is still used in some places; with modern equipment and medications, patients don't fight it the way they did in 1948. Still, I would never want it to happen to me.

A less severe but probably more dangerous treatment was insulin shock. Patients were given dosages of insulin to make them convulse and then quickly brought back again. After the treatment, we took them outside to the lawn and gave them a picnic lunch. Most didn't seem to know what had happened. It seemed to be a less frightening shock than applying electricity to the brain. I don't know how the doctors determined which shock therapy to use.

The only thing I enjoyed during those two months was sitting in on group therapy sessions. I never knew what they would talk about. Some things were pure fantasy, but sometimes someone would drop her guard and let us glimpse into the chaos of her mind.

Strangely, the only place I felt comfortable was with the men patients. I was five feet tall and weighed 98 pounds, so they could easily have hurt me. No one even threatened to do that. In fact, the bigger guys, the ones I should have been frightened of, were the most protective. They took their medicine from me without argument, and no fights ever occurred while I was on duty.

All in all, I hated that final two months of training and had nightmares for years afterwards of lying on wet cement with naked women fighting above me.

I was finished with formal training and ready to go home and take my exams. I never took part in a graduation ceremony, but that no longer mattered. I passed my exams and became a registered nurse wearing an angel's outfit: white shoes, white stockings, white dress, and a white cap. I regret that registered nurses no longer wear white uniforms. They don't even wear the white cap that signified where you trained. Dressing in white made me feel special.

Today a nurse aide, licensed vocational nurse, registered nurse, physician's assistant and even some physicians wear scrub pants and tops. Their only distinction is the color of the scrubs, which the staff understands but is a mystery to patients. Oh yes, their name tag states their position, if a patient can read it. I am thankful that did not happen until after I retired from my nursing career.

It took me four years instead of thirty months to finish training. My tuberculosis was in remission; I was twenty-one, and I had a wonderful little girl. I felt proud when I looked up at the sky and said, "Peggy, I beat the odds again!"

CHAPTER 5

Tumultuous Years (1949—1955)

MY SISTER FINISHED college and became a secretary at the U.S. Embassy in Bangkok, Thailand. While she was forging her path to success, I was starting on a path to destruction. It began with the stigma placed on unwed mothers. Miss Hudson's warning about decent men not wanting to raise someone else's child was like a record stuck on one song.

My self-image was very low when I met Jim. His large family lived near Dad. His father, who was Irish and Scottish descent, returned from World War I sick from mustard gas poisoning that permanently damaged the lungs of many veterans. He was much older than his wife who was born in Ireland and emigrated to the States as a teenager. They married and had six children in quick succession.

I don't know how he supported his wife and children before Prohibition created the bootlegging business. He learned how to make good bathtub whiskey and established a clientele. Before lung disease killed him at age 42, he taught his wife how to make it so she could support her children and stay home.

Through the bootlegging circuit, she met her second husband. They added four boys, making a total of 10 children she raised. She never drank the booze she was making. When Prohibition ended in 1933, most bootleggers dismantled their stills. Because some people were addicted to the potency of bootlegger's liquor, she and her husband continued making and selling it in smaller quantities until he died.

She left the four younger boys in the care of their stepsister, while she worked as a cook at a combination restaurant/bar. Here she met and married her third husband, who was the bartender. After that, she began drinking. She was never an alcoholic, but she was not a happy drunk.

Jim, the youngest of her first family, was bright and especially good at mathematics, yet never went beyond the eighth grade. When high school started, there was no money to buy him new pants. His sister had been given a white sailor suit, so his mother decided he could wear those pants. He was too embarrassed to go to school wearing girl's pants, so he hid in the woods. Liquor was readily available and made the days pass quicker. No one missed him because he had never reported to high school. Someone told his mother during Christmas break. By then he was 14, the minimal age for Roosevelt's (CCC) Civilian Conservation Corp.

When President Franklin D. Roosevelt took office, one of the first things he did was get Congress to agree to his CCC plan that gave three million young boys a chance to work, help their family, and learn a trade. The boys, who lived together in barracks all across the country, did not get to choose what trade they would learn. They received three meals a day (something some had never had before), and they each received $30 a month, $25 of which was sent to their parents to help support their families.

The CCC boys did more good than they were given credit for. They built new roads, strung electric lines in rural areas, constructed buildings and trails in national parks, and replaced or planted three billion trees between 1933 and 1942 when the program was disbanded. It also brought crime in the streets way down and helped many families provide for their younger children.

Jim's older brothers had enlisted in the service. Jim didn't like being a tree climber and landscaper. It was a good trade, but he feared heights and was terrified of being electrocuted while clearing limbs around electric lines. As soon as he turned 17, he joined the service. When he came back from the war, he married Sally in a formal Catholic wedding. Their marriage lasted only a brief time, and they had no children.

When I returned to my dad's cottage as a licensed R.N., Jim and I started dating. Prejudice between the English and Irish still existed, so what began as distrust between Dad and Jim turned eventually to hatred. Jim tried to keep Dad out of my life, and Gladys wanted to keep Dad's two remaining children out of their life. Dad and Gladys were married in a private ceremony. With her oldest daughter and a disabled mother living with them, they needed a bigger house and moved to a nearby town.

Dad's cottage was empty, so I stayed there. Sherry was still boarded with the same family as Gladys' daughter. Dad brought both girls home every weekend.

Jim's drinking was at the social level. He seldom drank at home; he wanted the social aspects, the comradeship he felt at bars. Maybe he needed the same feeling of belonging to a group that I felt. I went to smoky barrooms with him to watch wrestling on TV. Barrooms and rich people were the only ones that had TV at that time. Just as I had made the decision not to smoke, now I made the decision to stick with Cokes. Two Cokes would last me all evening, and that certainly wasn't true for his drinks. I was afraid I'd get addicted to alcohol when I saw how easily it happened.

At first, it didn't bother Jim or his family that I only drank Coke, but later they made hurtful remarks about me being "too good to drink with them," and "who is she to act so high and mighty." I wasn't impressed with wrestling, and the barroom atmosphere held no allure, so I often stayed home.

Because our relationship had become serious, Dad's prejudice about the Irish kept me from giving his advice about Jim being an alcoholic any credence. I was convinced that having an illegitimate child made my chance of finding a man without any problems non-existent. I could live with Jim's drinking because we were in love, and he always went to work even after a night of drinking.

Jim arranged for us to be married by a justice in New York he picked from the phone book. A church wedding was impossible because his first wedding was in the Catholic church. We couldn't afford it anyway. I didn't even buy a special dress. I wore my best dress, which happened to be black.

When we got to the justice of the peace, we discovered he was also an undertaker. We were married in the undertaker's parlor with his wife and an employee as witnesses. If I were superstitious, I might think the marriage was doomed. It was doomed, but not because an undertaker married us with display caskets as silent witnesses.

Sherry came back to live with us and Jim rented a room with kitchen privileges for us in Quincy, where there was a direct commute to Boston where I was doing private duty. The landlady took care of Sherry when Jim wasn't there. His drinking had gradually increased. I still believed getting him away from the drinkers in his family was all he needed. I actually thought *I* was capable of changing him!

Because he drank at bars, what I resented most was being left alone and having our paychecks disappear down a men's room urinal. Still, we were in love and I was happy.

Everything changed in January 1950. As on most Friday nights, Jim was drinking with his buddies. I was scheduled to work the next day, so I stayed home and went to bed early.

It seems Jim and his buddies drove to a bar in Boston later. Jim suddenly realized he was the only one without his wife, so he borrowed someone's car and drove home to get me. He didn't have a driver's license and was drunk. He woke me and insisted I go back to Boston with him. I realized the bar would be closing by the time he got there, but the car he had was his friends' transportation.

It was one of those impulsive decisions Miss Stratton warned me about. I didn't know how to drive, but I could watch for hazards ahead. Sherry was asleep, and I didn't want to get her up in the middle of a cold January night. My landlady's bedroom was across the hall. Even though it was late, I told her I had to go out but should be back within two hours. I left our door open. If Sherry woke up, the landlady would hear her. Chances are, she'd still be sound asleep when we got back.

Jim had trouble staying in his lane. I saw lights coming towards us faster than usual and urged Jim to move further to the right. I guess he was confused because he moved to his left, straddling the middle line. The other driver, who was more engrossed with his date than driving, was also close to the center. Before I could scream, the two cars sideswiped.

Jim lost control, skidded across a lane of oncoming traffic and into a telephone pole on the opposite side of the road. I remember the sudden impact of my head colliding with the dashboard. My legs were pinned under something heavy. As if from a distance, I heard Jim say, "Get out and run" as he took off into the darkness.

I drifted in and out of consciousness. I remember bits and pieces of action around me. Jim came back but then was gone again. Strangers were talking about my legs. The pain kept things from making sense, but one thing registered. "If we cut them off, we can get her out."

They were talking about me! I was screaming inside my head, "No! No!" but I don't know if any words came out. The last thing I remember was a voice saying, "I'm giving you this for the pain."

I woke up with full consciousness in a hospital bed with my right leg in a cast from toes to groin. How long ago was the accident? How did they get me out of the car when the motor had me pinned? Why wasn't Jim here? He must be taking care of Sherry.

A nurse told me I had been to surgery. The left leg wasn't in a cast, but it hurt to move it, and I never had a headache as bad as this one. She gave me a pain shot and blessed sleep took over.

A doctor woke me up. He told me I was probably 10 weeks pregnant. I thought he was wrong because I had no morning sickness. However an emesis basin cradling my face indicated I had vomited. He said the baby seemed to be fine. What about Sherry? Where was she? Neither the doctor or nurses knew anything about her.

My next memory was of a policeman standing over me asking questions I couldn't answer. At least he had answers for mine. Jim was in jail on charges of driving without a license and driving drunk. His mother had custody of Sherry. The other involved driver fled the scene but returned later. He was with a married friend and didn't want to get her in trouble so had taken her home. I was the only one injured, but the borrowed car was a total loss. Only the passenger side and motor collided with the telephone pole. I don't think anyone notified my father. He never came to visit.

Another doctor came with more information. He had not done the surgery but said I was extremely lucky. Their top orthopedic surgeon was on call that night. "Looking at the x-rays of your shattered right leg, we thought we'd have to amputate it. The surgeon insisted you were too young. He spent hours putting the pieces of bone back together with steel pins and rods. It may take a while, but eventually you'll walk again."

I never saw the surgeon who saved my leg. I was released in a cast and on crutches. As far as I know, I was never tested for a concussion or neck injury, but neck pain grew persistently worse over the years. My x-rays today show a misalignment of my neck *and* spine, which accounts for daily pain now that arthritis complicates it.

Jim was out of jail and working when I was released from the hospital. His brother brought me to his mother's house. We stayed there until Jim found a furnished house. This one was across the street from his sister and her family.

Insurance companies paid my hospital bills as well as for the totaled car. I would be in a cast six months and on crutches for another six months. Working was out of the question; yet it sometimes upset Jim, and I felt I wasn't carrying my share of the load.

Living across from Elaine helped. She drank too, but only in the evenings. It was good to have someone to talk with because Jim seldom came home until late. Throughout our marriage, my only social life was with his family. He never brought friends home, and I didn't have any to bring home.

The last week in July, I began dilating pains. My due date was still two weeks away, but the doctor wanted me out of the cast before I went into full labor. My first son was born on the first day of August. The accident, surgery, and pain medication seemed to have no effect on him. We called him Skip—the nickname his grandfather used. I walked with crutches or holding onto Skip's carriage for the next six months.

Jim bought a Christmas present for Sherry and Skip. By the end of January 1951, I could walk without assistance. Jim had managed to buy an old car and I assume he had a license. He seemed happy to no longer depend on others for rides. I was about to ask Elaine if she would babysit if I went to work when I discovered I was pregnant *again*.

Jim was gone more than he was home. He was still drinking but seldom came home drunk. In spite of my usual nausea and vomiting, I was happy.

On Easter, Jim dropped me and the children off for Easter dinner at Grammy's house. He said he had to go on an important errand. I thought he was getting Sherry an Easter basket and a toy for Skip. Maybe he'd get me a chocolate bar. However, he was gone such a long time I decided he was drinking. Grammy was carving the ham when I heard him drive up.

He walked in with Sally, *his first wife*! He announced they had been seeing each other since last summer and that they were going back together!

I was stunned. Why didn't he tell me at home when we were by ourselves? Why did he bring me here? Was he trying to humiliate me or was it a thoughtless blunder? He had not looked at me since his return. Had he been in love with Sally all the time?

Just as astonishing was the way the family, who never said a good word about Sally, welcomed her to dinner and invited her to sit where I usually sat. I retreated to a bedroom. It was as if I no longer existed in spite of Sherry sitting there as a reminder.

I heard them laughing and talking about old times. Was this the same Sally they professed to hate? What did the family say about *me* when I wasn't around? I never heard my name mentioned. I sat holding eight-month-old Skip and crying until Sherry finished eating and came looking for me.

I waited for Jim to come and talk to me, but he never did. I heard him leave with Sally. My eyes were red from crying, but if anyone noticed they were kind enough not to mention it. "Will someone please drive me home?"

Grammy said, "Jim wants you to stay here until he figures out what to do."

"Everything the children have is at home. I'm going home. Will someone drive me, or do I have to call my father?"

Grammy's husband drove me home. Except for Sherry's occasional chatter, silence surrounded us. Normally we didn't lock our doors, but for no logical reason I decided to lock both doors.

After I put the children to bed, I sat in the darkening living room thinking. I thought how little we really know about the people we love and think love us. I remembered when we first started dating, I overheard him tell a friend about me. He said, "She's not much to look at but she's a really nice person." Maybe I was more homely than I thought. Maybe that's why I had so much trouble making friends. When your self-esteem is low, it doesn't take much to hit zero.

It was my own low self-esteem that allowed Jim to make me a victim from the beginning. He deliberately tried to isolate me from my family and, with Gladys helping, was quite successful. The accident and pregnancies kept me from working, so I was totally dependent on him for even grocery money.

This final act of disrespect for me as a person was emotional abuse. I was a perfect target. Miss Hudson's reminders of the results of sinning still made me feel I had little value as a person. This cruel act pushed my self-worth off the chart. The crazy thing was I still loved him and wanted him back.

I was still sitting in the dark when I heard him trying to open the front door. Instinctively, I got up to open it when I heard voices. Sally was with him! Was that why he wanted me to stay at his mother's? Then he was at the back door. I heard him swear. I heard him say there was a window he could get in. I went back to the living room and waited with my hand on the light switch. He got the window open and was hoisting himself over the ledge when I switched on the light. He hung there for several seconds in disbelief before he began cussing me.

Sally was pulling on his leg, urging, "Let's go."

As his legs hit the ground, he shouted, "Damn you! You'll be sorry!"

I left the doors locked. Several days went by while I waited for word from him. I needed grocery money. Then one of his brothers came to get me and the kids. He told me Jim cancelled the rent on the house, which it turned out was two months in arrears. The furniture belonged to the landlord. It didn't take long to load Skip's carriage, our clothes, a few groceries, and Sherry's toys in the car.

"Where are you taking us," I asked.

"Jim arranged for you to stay at the house until you guys get your mess straightened out."

I was back living in a house where I knew I was not wanted. Grammy never showed her resentment when sober, and she had taken me in twice when I was stranded. Did she do it for me or for Jim? What I believed were her true feelings came out in hurtful remarks when she was drinking. I did not want to be there anymore than she wanted us dumped on her again—this time pregnant and with *two* children.

I told myself it would be straightened out when I saw Jim again. I couldn't believe he was serious about Sally after all the hateful things he had said about her. He was married to me. I was the one who gave him children.

When he came by a few days later to get something, he was sober. "I'm getting custody of Skip. Sally will raise him." His words were cold, emotionless, stating facts. He pushed me out of his way and left. Not once did he look directly at me.

I felt numb. Who was this man that looked like Jim but acted as if we were strangers? After he left, I realized he never mentioned the baby I was carrying. I asked a younger brother to watch the kids while I walked to town. It was a couple of miles each way, but by walking the railroad track I could cut the distance almost in half.

I went to the building where the priests live beside the church. A housekeeper answered the door and led me to a sitting room. I waited a long time before a priest appeared. He seemed uncomfortable as I explained my case. I'm not sure what I expected him to do, but I was shocked when he said, "Jim and Sally will always be married in the eyes of God. Your marriage is fraudulent. Jim has the right to have his son, and Sally will raise him properly."

"What about our little girl and the child I'm carrying?"

"Jim isn't the father of your illegitimate girl." I cringed at his choice of words, but he seemed not to notice. "That child is enough for you to take care of. As for the unborn baby, Jim doesn't think he is the father."

Jim had already been here! Why had he lied to the priest? He knew he was the father. The priest stood up to end my unwelcome visit. "Jim is only interested in his son who was never baptized because of you." He opened the door.

I fumed all the way home. Why did Jim only want Skip? Didn't he care about the baby I was carrying? Or was Sally unwilling to raise two of his children? I didn't know how religious Sally was, but Jim never went to church and had never mentioned baptizing Skip.

I felt totally abandoned. I didn't call my father because I didn't want to disrupt his new life. He had helped me more than I had a right to ask. He was so adamant about my not marrying Jim that I couldn't admit even to myself that he was right. I created this problem; it was up to me to solve it. Could I win in a custody battle if Jim told a judge the same things he told the priest? (In those days, there was no way to prove paternity.)

I was physically sick with my usual vomiting and heartsick about the possibility of losing Skip. What kind of person was Sally? All I heard was how badly she treated Jim, but was that true? Were his declarations of love for me all lies? How could I still be in love with him? Why did I still want to be with him? I could not make sense of anything.

One of his younger brothers told me Jim was trying to talk Sally into taking Skip. It seems she wanted no part of my children. I stopped worrying about a custody fight, but other problems remained unsolved. I had no doctor and no prenatal care. This isolated house was a long way from town. I had neither transportation nor a babysitter, so getting a job was impossible.

Jim showed up for Skip's first birthday in August and told me he had broken up with Sally for good. That great romance didn't last long, I thought vindictively. I asked what he planned to do about us.

"Ask me in two weeks" was all he would say.

Two weeks later *his mother* said, "Jim joined the air force. He's confused. He needed to get away by himself. He'll send me money for Skip and your unborn child."

"I'm their mother. Why doesn't he pay me?"

"Because you're living under my roof and I'm feeding all of you," she said bluntly.

When I went into labor on October 26, there was no one around. The younger boys were gone, and I didn't know when they'd be back. The older boys were married or out of touch with the family. Grammy and her husband,Tom, were working but assured me they would be home when the restaurant closed. Grammy said she would call Jim. I didn't know she knew how to reach him. I certainly didn't.

The only other person who would help me was my father, but it was hours before he got home from work. I couldn't go to the hospital unless I had someone to take care of Sherry and Skip. I might have to deliver this baby myself.

I got my scissors and a pair of shoe laces to cut and tie the umbilical cord and boiled them in a pan of water. While they were sterilizing, I put on a clean sheet, found clean towels and spread them on top of the sheet to keep from staining the mattress and put one of Skip's baby blankets close by. I put the scissors and shoelaces on a clean towel where I could reach them. It was not sterile conditions, but the best I could do.

I kept hoping someone would come home. I wasn't sure I could do this. Sherry's delivery resulted in a torn pelvis and Skip, even though smaller, required forceps to deliver him. My contractions were stronger and closer and nobody came.

Finally, I called my father. I was ready to hang up when Gladys picked up the phone. I briefly explained my needs. Dad must have broken all the speed laws. He carried Skip, and Sherry held his hand as I struggled into the back seat. He drove at break-neck speed to the hospital. I could tell by his silence that he was angry.

Half an hour later, I was holding another son in my arms. Jim never mentioned a name, so I called him Scott because I liked it, and it went with Skip.

After Dad dropped me at the hospital, he took the kids to the restaurant/bar where Grammy worked. Grammy, Tom, and a couple of her younger boys were leisurely eating supper. They claimed they thought it would be hours before I needed to go to the hospital and were waiting for Jim to get here. The truth was Grammy delivered her children at home with either a neighbor or no help and didn't think delivering your baby at home was a big deal.

At that moment, Jim walked in. Dad tossed Skip at him along with choice words about lack of responsibility.

Jim joined the air force to escape the chaos he created. Maybe he met Sally by accident and it brought back their passion. However, that didn't change his responsibility to me or his children. He returned to the air base while I was still in the hospital. He gave me no money; if I needed anything, I had to ask Grammy. I resented it and believe she did, too.

Scott was a colicky baby. That was no problem in the daytime because everyone was gone, but he cried a lot at night. Her comments to her husband were meant for me to hear. "Doesn't that baby ever shut up?" and "Why doesn't she take care of him?"

I was walking the floor, juggling him, and using a nipple pacifier. Many nights the only sleep I got was sitting in a rocking chair holding him against my shoulder. If I had a pediatrician, he would have advised a different formula. I had neither the money nor the means to get to a doctor. In the day, I was up with the other two children, so I existed on short naps. Even as young as she was, Sherry helped me.

I never told my father about my situation, so when he knocked on the door Christmas Eve, I was surprised. He brought toys for Sherry and Skip and hid them outside. He refused to come in so we talked outside.

"After the holidays, I'm going to work. There are places that hire nurses and have dormitories for them to live in." I didn't tell him the only ones I knew about were tuberculariums. That would have upset him.

"What about the children?"

"I've been checking the newspapers for local people who will board children 24 hours. I hate to ask, but is there any way you could lend me some money until I get a pay check?"

"Isn't Jim sending you anything?"

"I guess he's paying Grammy. He wants me to stay here and not work. I have to get out of here."

Dad swore—something he seldom did. He agreed to give me money to get the boys established and said he would place Sherry with his stepdaughter again and pay her board until I got my life straightened out. He suspected Jim would not support her. He was right.

I was emotionally and physically exhausted. Working at a tubercularium was the only way I knew to start earning my own money again. I refused to think of the danger. I wanted to believe that first time was a mistake. Otherwise, how would I have been released in six months? (*What I did not know then was that my TB was first caught in the "latent" stage. It could be seen on the x-ray but had not reached the active stage. The bacteria were still there and could become active at any time.*)

This tubercularium had a dormitory for its nurses with free meals and no transportation worry. Where else would I find that? Besides, I liked working with long-term patients.

I found a lady through a newspaper ad who was looking for a boy her son's age to be his playmate. Skip was a perfect match, so although she had only planned to babysit days, she was willing to board him 24 hours a day. She needed the money. It was with great reluctance that she later agreed to take Scott when his first babysitter didn't work out.

I started work the first week of January 1952. Now I had three children that I couldn't have with me, but the money I earned was mine, so I began a savings account. My goal was to get an apartment near a hospital where I'd make more money and only need a babysitter when I was working.

A year after Jim left me for Sally, he came home on leave and took me to supper. He said when his service commitment was up, we would go back together and that he wanted to adopt Sherry. We had discussed this when we got married, but something always interfered. I agreed because I thought it was important for all my children to have the same last name as me. How would I explain why she was using my maiden name?

Jim said, "I'll pay the boys' board, but since you are working you'll have more money than me. You should buy their clothes and pay for Sherry." (I didn't tell him my father was doing that.)

I thought my life was finally getting back on track. A month later, I was proven wrong. I had been losing weight and the staff doctor noticed it. I had had an x-ray when I started work and wasn't due for another until the first of May, but the staff doctor insisted. When he compared the new x-ray to the January one, the difference was undeniable. This time I saw the proof on x-rays with my name on them.

I had learned more about TB now that I was working with patients whose disease was active. I already had two of the symptoms of active disease: fatigue and weight loss. I wanted to believe they were because of the stress I had been living with for a year, but stress can also cause latent TB to become active.

I was immediately admitted to the tubercularium where I'd been working. Two new drugs that had just become available gave me hope. Paraminosalicylicacid (PAS) was a powder dissolved in water that tasted awful. I took it after each meal. The second drug was streptomycin, which was injected deep into the buttocks three times a week. It was a painful injection.

For some reason, I have few memories of my year there and didn't make any friends. I had been in isolation in a private room for six months; after I was moved to an open ward I didn't want people to know how much Jim drank. He only had leave a few times, but I never knew when he would arrive drunk.

His arriving drunk happened late at night when Jim came to "take me home." He must have been too drunk to reason with because both the doctor and police were called. The doctor told the police that I was the only one who could legally make the decision to leave against medical advice.

The doctor and a policeman woke me up. "Your husband is demanding to take you home. He is very drunk. If treatment is interrupted, the bacteria can become active again and will grow back in a drug-resistant form. If you go with him, you'll be signing yourself out against advice, and we will *not* take you back," the doctor warned. Knowing how unpredictable Jim was when drunk, I said I did not want to go with him. "Then go back to sleep. Don't worry. We'll handle this," he said.

I've never known what happened. Did the police take him to jail? I doubt wearing a uniform made any difference if he was belligerent. He never came back to visit again. He may have been blacklisted, he may have been too embarrassed, or he may have blamed me for whatever happened.

The holidays of 1952 came and passed. Grammy and several of her sons moved to Florida where Elaine was living after her husband's death in a job-related accident. She received a large settlement and would never have to work but was depressed and needed emotional support.

Dad and Gladys started visiting me when they learned Jim wasn't. We never discussed him, staying with safer subjects. I had a surprise visit from my brother-in-law, George, who was being sent to Korea. He wanted to say goodbye to his family but found the house vacant.

He located me through one of Jim's bar buddies who explained Jim was in the service and most of the family had moved to Florida. George had not heard from anyone since he joined the army months before. I told him what little I knew. I felt sorry for him, and he must have felt sorry for me. When he left he wrote down a phone number. "If you need help, call that number. He'll know where I am."

I had started my second year at the tubercularium with no release date in sight when a nurse brought me a letter saying the state was taking my boys and putting them in foster care. It seems Jim had not paid for their care for several months and the babysitter's pleas went unanswered. I didn't blame her. She was taking care of our boys because their family needed the money.

I had paid back my father and paid for clothes and toys for the kids. I had no money left. Jim got money from the government. Why wasn't he paying the board? I was trying to decide what to do when Jim arrived. He had received the same notice and got emergency leave.

This was the first time I had seen or heard from him since his drunken night visit. The date for the state to take the boys was only a couple of days away, so we had to act fast. I signed myself out in late July 1953. The doctor tried his best to make me stay, but I had to get my children, and I couldn't trust where Jim would put them.

The doctor's parting words were, "Without the medicine and care you get here, *you'll be dead within five years*."

I thought of Peggy's words, "Take it one trick at a time." My immediate problem was to get my boys before the state took them. Jim had a friend with a car waiting outside. I quickly gathered my belongings throwing them in bags a nurse gave me. I didn't answer the questions coming from all sides.

When we arrived, our babysitter started crying. "You have no right to take the children or their things until you pay me."

Jim ignored her as he loaded the kids into the car with the clothes they had on. I was trying to apologize when Jim grabbed my arm and pulled me to the car. He wanted to get out of there before any police arrived.

We had nowhere to go except Grammy's. She had left behind furniture, linens, blankets plus some dishes and some kitchen utensils. I cleaned the house while Jim's friend drove him to a store to get food and have the electricity turned on. That meant he did have some money. How much I do not know.

Things looked presentable when the state social worker arrived two days later to check where we were living. She said even though we owed money, our babysitter could not keep the children's clothes. She had already been there and had their clothes in her car. She also had a baby carriage that belonged to us. Returning children's essentials did not apply to their toys, which could be held until we paid the money we owed. We never got them back.

Sherry stayed at the house where she was boarded through the rest of the summer and the next school year because there was no bus service from Grammy's house. Dad continued paying her board.

Jim sent me money once a month to buy groceries and heating/cooking oil. He was getting money all along, so there was no reason for not paying the board. I couldn't work without transportation or a babysitter, so I was grateful to get money, even though it wasn't enough. I couldn't leave the boys alone, and the walk to town was too far for them, so we took a taxi to shop once a month. It took careful planning to have something left before the next money came.

Jim came home on leave in October and found the house was cold with no coal for our furnace. After his visit, an unexpected load of coal was delivered. In order to get the furnace started, I needed wood. So unless it was snowing, I bundled the boys up each day and we walked to the nearby ocean inlet where there was always driftwood after the tide went out.

Scott rode over in the baby carriage without its mattress but had to walk home because the carriage was loaded with driftwood. It was hard to push the carriage over the rough ground, but I couldn't think of any other way to get the wood home. I spread the new wood on the basement floor to dry and used the wood picked up a day or two earlier to restart the furnace.

When the electricity was turned off in January because I couldn't pay the bill, I made more changes. The first few days after the monthly check arrived, we had fresh hamburger and hot dogs. After that it would be canned tuna, canned Spam, cheese, eggs and milk for protein to go with canned vegetables. I used evaporated canned milk and diluted it with water or powdered milk that I mixed with water as we used it.

I made potato soup with an onion in it, and on Fridays added a can of corn and called it corn chowder. We had canned beans and potatoes a couple of times a week with canned brown bread that had raisins in it. We had spaghetti with a can of tomato soup.

We had oatmeal or corn flakes for breakfast and peanut butter and jelly sandwiches until we ran out of bread. We had scrambled eggs for breakfast or supper. Every Sunday, as a special treat, we shared a can of some kind of fruit. The boys, being so little, didn't eat much, and neither did I. We never missed a meal of some kind.

We always had kerosene for our kitchen stove which was both our cooking and heating source. The oil company filled the outside tank once a month and I filled the five-gallon inside bottle from that. Electricity was a luxury we could do without, but we needed kerosene.

I devised "a cave" we lived in during the day. The boys loved the idea. I hung a blanket as a barrier from the kitchen entry (it had no door) to the rest of the house. We kept the door from kitchen to the back porch shut. The enclosed porch wasn't insulated, so it was cold. The refrigerator might seem useless without electricity, but it was a place to store bread, eggs, margarine and powdered milk. We lived in the kitchen cave during the day. We left the oven door open to provide more heat. On the coldest days, we wore an extra layer of clothes.

We had cold water at the sink and bathroom, but I always kept the big tea kettle on the stove filled with water. When we needed extra hot water, I filled pots and heated them on the stove, too. I washed dishes, clothes, and our bodies in the kitchen sink.

To dry clothes, I hung a rope across the kitchen from one corner to another. There were always clothes drying, even though we only changed when it was necessary. Thankfully, Scott was toilet-trained, so there were no diapers to wash.

Using the toilet was a challenge because the bathroom was cold. We made as few trips as possible into any part of the house closed off by the blanket curtain to our cave.

The boys played with matchboxes, bottle caps, and clothespins which they pretended were trucks and soldiers. I read the same book to them over and over. On sunny days, they went outside to get fresh air and burn off some energy.

Dark came early, especially on cloudy or snowy days, so we used a kerosene lamp for light in the kitchen and a flashlight to go to the bathroom. This was also the time to "fix the whiskey bottles." I had salvaged several empty bottles from a dump behind the house that I filled with heated water and then placed them in the oven till they were really hot. Then I lined the bottom sheet with them to warm it, as well as the blankets on top, before we went to bed.

We slept together, gaining extra warmth from each other. We wore two pairs of socks, ski pants, and a sweater over pajamas. Little children are so adaptable, and I learned to be flexible.

I had severe morning sickness from January until May. I had been warned *not* to get pregnant by the TB doctor, but birth control of any kind was not available in Massachusetts. The one doctor I saw after leaving the tubercularium advised me to go out of state to get a vaginal protection device, but I had no way to do that. I don't know if men could still buy condoms as they could during the war. Certainly, in the air force Jim had access to them, but he refused to use them.

My spirits lifted when spring came. Jim had a hardship discharge. He said he was trying to get it for some time. He used my pregnancy, tuberculosis, and three little children as a reason.

In May things were better. I was over morning sickness. Jim paid the overdue electric bill and we had electricity. We no longer lived in the kitchen cave. Jim was drinking, and I never knew what kind of a mood he would be in when he got home. He talked constantly about moving to Florida where his mother and most of his family were. With the school year over, we brought Sherry back into the family.

In late July, I received word my grandfather had died in a nursing home, and I was the executer of his will. I knew nothing about the responsibilities of executers, and I was eight months pregnant. Jim was gone more than he was home. My grandparents' chicken farm had been closed down several years before when my grandmother died.

The house had to be prepared for sale. My sister, who was still in Thailand, and I were the sole heirs. I told Jim I was going to stay there. It was the only way I could prepare the house for sale. He didn't like it because it wasn't within walking distance of any bars, and he had not bought another car after getting out of the air force.

I had the telephone connected and the electricity turned on, although we could get by without it. The house had no air conditioner or fans. The "refrigerator" was an old-fashioned icebox with ice delivered. The cost for the electricity I used was minimal and allowed me to use Grandma's electric washing machine. It was much easier than washing in a sink with a scrub board. More important, I no longer had to wash Jim's landscaping clothes in the bathtub using a toilet plunger to get them clean.

Living here brought back memories. On the landing at the top of the stairs sat an old cedar chest. The only bathroom was upstairs, so when we visited our grandparents we walked past that cedar chest frequently. I wondered what it contained, but it was kept locked. I know because I always tried to lift the lid. When I asked Grandma what was inside, she said, "My treasures."

Now I had the key. My hand quivered with the excitement of a child promised a surprise. What a disappointment. All I saw was linen sheets and pillowcases still in cellophane wrappers. Yet on her beds, even now, the sheets and pillow cases were chicken feed sacks she had washed and sewn together. Why had she never used these sheets?

I tossed them on the floor and found layers of towels that were never unfolded. Where were the treasures? I dug deeper until my hand touched a box. Pulling it out, big letters on the top announced, "Sears and Roebuck."

A memory of long ago overwhelmed me. I sat on the top step, holding the box to my chest, remembering the time I was in a school play. The other girls were getting new dresses. When I asked if I could have a new dress, Mama had tears in her eyes. "I'm sorry. Grandma's birthday is coming, and we're saving to buy her the waffle iron she keeps talking about."

I opened the box. There it was in pristine condition with warranty tags still attached. I cried a long time. I wasn't crying for a little girl who didn't get a dress she wanted. I was crying for my grandmother who never used any of her treasures. What was she waiting for? I was crying because, like my mother, she died without ever having really lived. And I was crying because it looked as if I would die as unfulfilled as they did.

The threat of tuberculosis was constantly with me along with the doctor's prediction of my death. I knew Jim was not dependable, but he loved his children. I told myself he would surely stop drinking to take care of them when I died. But I wasn't ready to die yet. I had barely used the nursing career I struggled so long to obtain. My life had to have more meaning than Grandma's cedar chest full of stuff.

Now I had another baby on the way and I couldn't even take care of the three I already had. Whether I liked it or not, I needed Jim. My sister and I had inherited this house, and I could have stayed here. But, like Grammy's old house, it was isolated with only old people for neighbors and no transportation within walking distance. I had no car and didn't know how to drive. And, for some crazy reason, I was still in love with Jim. My marriage vows would need more challenges before I was ready to give up.

Two of my grandparents' neighbors arrived to ask for the special chinaware they had been promised at my grandfather's death. (*I learned years later that these were valuable antique dishes.*) With little kids, you use cheap dishes. My sister may have wanted them, but she was in Thailand. I told neighbors to take whatever dishes they wanted. They also took most of the smaller items like linens, towels, and that waffle iron that was never used. A few hours later, another neighbor came.

"I heard you were giving away Mabel's things. Can I have the chime clock in the den?" She also took the last of the special chinaware plus the china cabinet itself. It was one less thing to get rid of. I did not know "old" furniture was valuable.

August was a busy month cleaning the house and discarding junk. Since his discharge, Jim had been working regularly—although not always on the same job. He gave me what I needed for food and utilities even though he still spent a lot of time and money at bars. Somehow he always found someone willing to bring him home.

In Massachusetts, we never paid attention to hurricanes. They happened down south. I didn't know anything about them until Hurricane Carol hit the Massachusetts coast. We awoke to a nasty day, raining and windy. Jim went to work thinking they'd have to pay him for showing up and would then send him home.

He called to tell me about the hurricane. "Don't worry. We are far enough inland so there should be no damage. Turn the radio on. If it gets really windy, they said to take the kids down to the basement until it dies down. They won't let me come home. Our company may have to help clear trees. I'll get extra pay."

I turned on the radio and was stunned to hear a worried voice advising people to seek shelter immediately. To be safe, I decided we should go to the basement. I didn't see how the wind could get any stronger than the gusts that made the windows rattle. I made sandwiches and then took the children to the old egg-sorting room. I came up and got some milk and glasses along with a big blanket and pillows because the floor wasn't very comfortable and I had no idea how long we would be there. Our radio was too big and heavy for me to carry to the cellar. To make matters worse, I began labor contractions. Maybe it was false labor due to stress. Sometimes that happens.

It seemed we were in the basement for hours before the wind and rain died down. This homestead had been cleared for a chicken farm, so there were few trees. When I came back upstairs I was amazed to see one big tree near the house had fallen away from the house. Had it fallen toward the house it could have gone through the roof. I went outside to check. We lost several trees on the hill where I played as a child. The roof on the old barn was gone—completely disappeared. New owners would probably tear it down.

When I came back inside, I noticed how quiet it was. The radio was making no sound. I tried a light switch. We had no electricity. The phone rang. It was Jim calling to see if we were all right. I told him my labor had started. He promised to call Carol and Jerry, who were my transportation and babysitters. While we were still talking, the phone went dead.

For supper, I gave the children cold cereal and milk. I was making preparations to sterilize the things I might need, when I heard a car and ran to the window. It had to be Jerry and Carol. It was a taxi.

George got out. I ran to the door to ask the driver to wait. George had already done that because he wasn't sure this was the right house. He had just been discharged, and Jim was the only connection he could find to his mother. He had gone to one of Jim's favorite bars and got directions. It seems one of the steady customers had driven Jim home a couple of times.

I asked George to stay overnight in case the babysitters couldn't get here. *(I later learned that Jerry and Carol didn't arrive until late the next day. Jim had been drinking when he got home, and instead of thanking George, he accused him of having slept with me. I believe it was the last time the half-brothers ever spoke to each other.)*

As soon as I was in the back seat of the taxi, the impatient driver took off. It was a thrill ride because trees blocked the roads in places, which meant turning around and finding another route or going off the road onto someone's land to get around a fallen tree. The taxi driver was sweating (and swearing at every delay). He kept telling me not to have the baby yet. He kept watching me in the rearview mirror, so I tried not to groan with the contractions. He ran to get a nurse when we pulled up in front of the emergency room and took off before I could pay him. I wasn't sure whether George had paid him either.

It was a couple of hours before I had the baby. The emergency room staff was upset because I had no doctor and no prenatal care. I knew whatever doctor was on call would have to deliver me just as they had done for Scott. Of course that wasn't the right way, but Jim thought it was a waste of money to pay a doctor to say you are pregnant when you already knew it.

Going to a hospital to have a baby was expensive enough. He never understood why I wanted a strange man to deliver me. He allowed me to have prenatal care and a hospital delivery for Skip because I was in a cast from the car accident. For the other children, he reflected his mother's viewpoint that women should be able to deliver at home.

The labor room was full, and the nurses were busy. When it was time, they wheeled me into the delivery room. Then a nurse was holding my legs together and telling me *NOT* to push. This baby was demanding to get out. Cathie was the only little girl on my ward born that day who was not named Carol after the hurricane. I named her Catherine after myself so I could call her Cathie. It was the name I wished people had called me instead of Kay.

I fell asleep as soon as I was moved to a bed. The next morning the doctor who delivered me introduced himself. He quizzed me about prior pregnancies *and* my tuberculosis. He scolded me for getting pregnant and not seeing a doctor during my pregnancy. I didn't try to explain that Jim made those decisions and only paid bills he thought were necessary. When I admitted I had three children at home, he insisted I have my tubes tied. "You aren't strong enough," he scolded. "If you get pregnant again, I guarantee it will kill you."

I thought he was exaggerating, but knowing how easily I get pregnant, I knew I couldn't take care of any more children. One of the few reasons a doctor could perform a tubal ligation in Massachusetts at that time was if pregnancy was dangerous for the woman. However, it also required the husband's signed approval or it would take a judge to override him. I knew my only hope was to get Jim's signed approval.

He visited once. I asked the nurse to call the doctor immediately. I pleaded with Jim to wait and talk to him, but Jim never liked hanging around hospitals and didn't want to talk to the doctor after I explained the reason for a conference. The doctor arrived as Jim was leaving and stopped him.

At first Jim refused to sign. Then he said, "I'll bring her back in a few weeks for the surgery."

"You'll never bring her back," the doctor replied.

I knew he was right. Jim would find excuses not to sign the papers, and I'd be pregnant again. The doctor persisted. "I won't release her until she has surgery. I promise it won't affect your sex life if that's what worries you, and your church will understand this is to protect her life. I'm surprised this pregnancy didn't kill her. Another one will."

This was the second time doctors told him another pregnancy would kill me. Maybe he believed it. Or maybe he only wanted to leave and get a drink. He signed the form. I would remain in the hospital two extra days. The electricity was back on, and I was anxious to go home, but I knew if it wasn't done now, I'd regret it.

When I got home after 12 days in the hospital, both Carol and Jerry were eager to leave. While she was getting her things, Jerry whispered, "Carol and Jim are having sex."

"What? Why do you think that?"

He shrugged his shoulders and walked away. I decided Jerry was drinking and gave it no more thought.

Jim admitted he never went back to work after that night they wouldn't let the crew leave. He had to clear telephone and power lines and swore he'd never again be in a position where he was forced to do that. He had borrowed enough money for us to go to Florida and insisted we leave immediately.

I couldn't do that. I was just out of the hospital, and I was responsible for getting this house ready to sell. I had a new baby. Sherry had just started school. I pleaded with him to wait until the next school break. We finally agreed he would leave now, and I would come as soon as I could.

"By then, I'll have a job and you won't have to work. You can stay home with the kids," he said. I took his promises with skepticism, but it gave me time to do what was necessary here and get some needed rest; I could sleep without worrying about him.

When the kids were settled down or sleeping, I went through drawers and prepared charity bags for anything that looked usable. I trashed the rest. Finally, when I was down to the old furniture, I called a secondhand dealer. "I'm cleaning out my grandparents' house with lots of old furniture. I'm looking for someone who will take everything."

He arrived 20 minutes later. I was impressed. "I know everything is old," I said, lifting a table cover to show him the round dinner table with a polished surface that didn't have a mark on it. Removing the cloth also exposed legs that had been carved years ago. "You can see it has been well taken care of." He did not respond. I saw him eyeing a wooden rocking chair with carvings.

We went into the parlor. Besides a piano, a parlor table, and two upholstered rocking chairs, there was a love seat of some exotic wood that had carvings on the top, arms, and legs, and needlepoint upholstery on backrest and seat. This captured his attention.

I confessed, "It is pretty, but uncomfortable to sit on. My grandparents only used it when the minister visited." He never replied. He was making me nervous. We walked through to the small sewing room with an old foot-pedal sewing machine.

Next was Grandpa's den. There was a carved table there that had been piled with old newspapers that I had thrown away. Now clear and polished, I thought it was attractive.

We were full circle back to the kitchen, and he had not said a word. We walked through the "cold room" where there was only the old-fashioned icebox and a sink for de-feathering and cleaning chickens. This led to the back porch that had a hammock, an old work table, and a couple of chairs.

"This is where I played when I spent summers here as a child," I said. The quiet was deafening. He showed almost no interest in anything. Maybe everything was too old. We went back through the kitchen to the front doorway where stairs led past the cedar chest on the landing to the bathroom and two bedrooms.

We walked through the bathroom with a big claw-foot tub and a pedestal sink with Grandpa's shaving cup and razor strap hanging on the wall above it. Grandpa's bedroom was smaller with an old-fashioned bed and matching dresser. "I like the carved designs on the headboards," I said in case he hadn't noticed them. Then we went to the larger bedroom. "This was Grandma's room." I felt as if I was talking to myself.

I wanted this to end. "What can you give me for everything in the house?"

He frowned. Then finally he said, "How does two-fifty sound?"

I hate negotiating. "I was hoping to get at least three hundred dollars."

He smiled for the first time. "Lady, I feel sorry for you. I'll give you three-hundred-and-fifty dollars if we make the deal right now."

Wow! He gave me the money in cash, and I signed his receipt. He put the parlor table and a rocking chair in his car and said he'd be back with a truck. After I gave my sister her share, I would have $175 cash, and I had got rid of everything. He even took Grandpa's shaving cup and razor strap.

Years later, a friend asked if he might have meant two-thousand-five-hundred when he said two-fifty. It seems even $2,500 was way underpriced. I had given away everything but the house and land, which were managed by a realtor. After I learned old furniture was considered valuable, I felt guilty for cheating my sister out of part of her inheritance. It didn't bother me because it would have been wasted on booze anyway. If Grandpa had lived a few more years and I had known more, that money could have changed my entire life.

I used my $175 to take the kids on a bus to Florida to what I hoped would be a new beginning to our marriage. I made sure the hospital bills for Cathie and my surgery would be paid from the house sale and the rest of my share sent to me in Florida.

Elaine arranged for us to live in the housing tract where she and her son lived. She was still having a rough time emotionally. It would be nice to have her for a neighbor again, and Grammy lived in a nearby town.

Sherry was in second grade. She had always liked school, but she hated the Florida school, which seemed to be teaching first grade again. Sherry said, "Some days we write our numbers from one to a hundred over and over and over. Why does she make us do that?" I had no answer.

When we moved into this furnished house, I didn't expect nice furniture. I'd been down the rental road before. Still, I was appalled at how filthy the house was. It was as if no one had cleaned it since it was built. My first task was to remove the rubbish strewn everywhere. While doing that, I was introduced to Florida's cockroaches (called palmetto bugs).

I'd had my share of nasty cockroaches. I hated them but didn't fear them because they run away when you put on a light. Florida's cockroaches were huge; they could *fly,* and instead of running away, they attacked me. They didn't bite, but it scared me when this creepy thing flew at me when I put a light on. They hid in dresser drawers, closets, and shoes.

About a week after I got my share of money from my grandparents' estate sale, Jim walked to town to get a quart of milk and didn't come back.

Two weeks later, Grammy told me he called her. He was back in Massachusetts visiting friends and would return to Florida in a month or so. He took the estate money but paid the month's rent that was due. He knew if I got desperate, his mother or Elaine would help me. He seemed devoid of any sense of personal responsibility toward me or his children. He was so different from my father that I couldn't understand it.

I looked for work when neither he nor the milk showed up in two days. The only opening at a nearby hospital was evening shift, but I had weekends off. I was at home most of the time with the younger kids while Sherry was in school. My shift was 2:30 to 11, so I was home before midnight. Then Elaine watched the children, fixed their supper, and, with Sherry's help, put them to bed.

Jim's "month or so" lasted all winter, spring, and most of the next summer. Soon another school year would start. I had no indication of when he was returning. Then he called Elaine and I was able to talk to him myself. He asked if I had enough money for bus tickets to Boston. He had decided to stay in Massachusetts and was buying us a house on his GI bill.

Sherry was delighted not to have to start third grade in Florida. I was happy that Jim was going to buy us our own house. I thought how nice it would have been to live at my grandparents' farm if I only had a car and knew how to drive. The beauty of growing up in the city was buses, trolleys, the subway system, and even taxies provided transportation everywhere. Many people never owned cars.

I worried about my life sentence now down to three years. When I died, the children would need Jim. Yet there was a nagging fear that they could not depend on him. Would he dump them on Grammy or expect a new wife to love and take care of them?

I tried to think of positive things as I packed up what I could in suitcases and boxes for the long bus trip. I kept telling myself that he had changed because we were going to have our own house. That must mean he wanted us to be a real family. Perhaps he had stopped drinking. There was always that hope.

I refused to think about the times he disappointed me.

CHAPTER 6

How Love Dies (1955—1961)

JIM AND JERRY met us at the bus station. Jim announced our VA loan would not go through for a couple of months, but in the meantime we would stay with Carol and Jerry. They had two unused bedrooms upstairs that would be ours, and we had full use of the kitchen. Disappointment was like an unexpected punch in the stomach. Why didn't he let us stay in Florida until our house was ready? It seemed as if we were always living with someone else, an invader of other people's privacy.

Their house was one of three on top of a hill. The people in those houses never spoke to each other. The agreement was "stay out of my business, and I'll stay out of yours." It seemed weird, but I kept my opinion to myself.

Since Carol didn't work, she would babysit my kids and her own son. It was obvious Jim expected me to work. Finding a job was easy, but because of our location, transportation to most hospitals required a long walk and bus changes.

The first job I found was an evening shift with every weekend off. I was going to turn it down and hope for a day job, but Jim seemed to think this shift was better. He was working and seemed pleased about my having weekends off. I soon realized it was Carol who wanted the weekends free. Since I was home, I could take care of her son, too.

I walked two miles and took two buses. It took two hours each way except during snowstorms when buses were late or I missed one.

We had one really severe storm that winter. By the time I got off work, snow on sidewalks was up to my knees and visibility so bad I feared walking in the partially cleared street. Nurses were offered a chance to work the night shift at overtime pay because night nurses were calling in to say they couldn't get to work. We had no home telephone; if I didn't come home, everyone would worry.

Getting home was a nightmare. Buses were late and walking was a slow, tiring exercise in willpower. When I finally made it up the hill, I found Jim in the kitchen drinking whiskey and playing poker with friends. The bar closed early because of the storm, but he was too drunk to remember. He demanded to know where I had been. Before I could answer, he said he knew how long it took to get home. His friends looked embarrassed; I was exhausted, so I snuck away when someone distracted him. He was obsessed that I was cheating on him. I could not understand why. He fell asleep at the table and didn't remember anything the next morning.

Not long after that, a male nurse joined the evening shift. He drove past the hill where we lived and was delighted to give me a ride in exchange for the bus fare. It cut my travel time from two hours to a half hour. I told Jim about the new arrangement when he was sober and explained it was a male nurse. He seemed okay with it since I got home much earlier.

The arrangement continued for a few months. The few months of waiting for his house loan to go through had turned into all winter, and now spring was here. We were still living with Carol and Jerry. Then one night as my driver and I were walking to his car, someone yelled, "Kay, Jim's over here!"

I thanked my friend and said, "I guess my husband is picking me up. See you tomorrow." When I reached the pickup, I saw Jim was very drunk. He was cussing his friend for warning me he was there. Jim got out and pushed me into the middle.

"Why didn't you follow them like I told you?" This friend knew Jim well enough not to argue, but I wish he had followed me home. Jim kept asking me who my lover was. No matter what I said, he hit me on the head with his left hand that rested on the back of the seat.

His friend said, "Kay, do you want to go into the police station?"

I suddenly realized we were in front of the police station. Jim didn't realize it either until we pulled up to the curb. Two policemen in front were watching us. Jim yelled at his friend to take off, but he didn't. An officer yanked Jim's door open. "Anything wrong here?"

"No," Jim snarled.

"She's his wife. I'm afraid he's going to beat her up."

The policeman asked me, "Do you want him arrested?"

"Yes," I said impulsively. That was a big mistake.

Jim was not a big man, but he was strong, and when he was drunk, he loved to fight. He never seemed to feel the pain of other people's punches. While three officers were trying to pull him out of the car, he was swinging at them. Another officer arrived, and it took all four of them to get him on the ground and handcuffed. As they dragged him away, he threatened me, his friend, and the police.

Another officer started questioning us. Jim's friend supplied the details. He found Jim drunk at a bar and offered to drive him home. When he made the offer, he didn't know it included going to the hospital to pick me up.

The officer told me to be sure and be in court in the morning when Jim would appear before the judge. He also said, "Looks like you picked the wrong man, lady."

In court, Jim never looked at me. The charges were threatening assault and resisting arrest. There was no reason for me to be there! The actual trial would be in a few weeks, so the judge released him with a warning but without bail. I decided I better wait for him by one of the courthouse doors. When he came out, he glared at me and then went in the opposite direction. I found a cab and went home. He did not follow me.

Carol was babysitting. When I told her what happened, she left immediately, saying she'd be back in time for me to go to work. I had time to do a load of clothes using her electric washing machine. These machines were a great luxury to me. I was ready to hang the wet clothes outside when I heard a car driving up. I expected it to be Carol. The older kids were due home from school soon. I had to hurry. It was nearly time to get ready for work.

I was on my way to the door with the clothes when Jim opened it. He had obviously been drinking, but there was something different. He had a strange look in his eyes and a twisted smile. "Think you're smart, huh?"

I didn't see the punch coming. It knocked me down, and the basket of clothes went skidding across the kitchen floor. I tried to get to my feet, but he was straddling me and yelling, "You'll never call the cops on me again, bitch."

I saw the second punch coming before it connected with my eye. It felt as if my eye had popped out of its socket. I thought it was gone and that scared me more than the pain. Crawling on my hands and knees, I tried to get away from the kicks raining on me.

Finally, I made it to the living room couch and curled up in one corner trying to protect my head and face from blows that punctuated every filthy thing he could think to say about me. Suddenly all the kids were there. I heard crying. Sherry must have taken Cathie upstairs. I don't remember seeing them again. I do remember Skip clinging to Jim's arm as it rose for another swing.

The room was spinning and then everything went black.

I was lying on the couch, half-conscious of what was happening. Jim was no longer hitting me. I heard him say he was sorry, but he was telling Skip, not me. Then I saw his face close to mine. He looked different and his voice was barely more than a whisper meant only for me to hear. "If you ever call the cops on me again, when I get out of jail, I'll kill you."

I don't think he hit me again, but I lost consciousness. When I woke, things were happening that didn't make sense. Carol and Jerry were home, and they were all arguing about something. I didn't see any of the children. Where were they?

Then Jim carried me to the car, and Jerry drove us to a doctor who kept asking me to do weird things like watch his finger move in different directions. He said I had a mild concussion and to rest for a few days. "If you start having headaches or faint *again* (?), call me. Don't worry. Your eye is still there. It's a black eye," he assured me.

A nurse kept checking my blood pressure while the doctor took Jim somewhere. They were gone a long time.

In the next few days, I did a lot of thinking. The neighbors must have heard me screaming. Why didn't someone call the police? But then, what good were the police? They might provide protection to end a particular episode, provided they arrived in time, but in the end they accelerated the problem. How do you protect yourself while waiting for a court hearing? Getting probation for a first offense was probably the worst that would have happened to Jim if I had gone back to court. Even today, our justice system is more concerned with the rights of the accused than their victims. The only knight in shining armor is in fairy tales.

Previous emotional and verbal abuse had finally become physical with threats of worse "next time." It was like the night I got pregnant: I went willingly, so getting pregnant was my fault. I married Jim stupidly thinking my love would stop his drinking. Now I had to learn how to avoid his drunken anger or be prepared for the consequences.

Jim never hit me again. He didn't have to. I was afraid of him when he was drunk. No matter what he said, I didn't respond or would say what he wanted to hear. He never seemed to remember the next day.

Carol didn't have a phone, so I couldn't call in sick. With my bruises, I couldn't work. I started to walk down the hill to tell my ride, but when I started to vomit, I knew I couldn't make it. I wondered how long he waited before he went to work.

I had not gone back to work and doubted that I still had a job when Carol, Jerry, and Jim came home late one night laughing and joking. They had been drinking together. Then Jim or Carol said something that upset Jerry. He blurted out, "I know you've been screwing my wife."

Jim turned around and yelled, "So what are you going to do about it, fat boy?"

He walked up to Jerry and started pushing him backwards toward the door, taunting him with every push. I realized Jerry was as scared of Jim as I was. When Jerry's back was trapped against the door, Jim said, "Why don't you be a man and fight for her, you coward?"

Jerry did not respond.

Jim hit him once in the face, his knees buckled, and he went down in a kneeling position. Carol grabbed Jim's arms and urged him to come into her bedroom so she could talk to him. I snuck upstairs, acknowledging to myself that I was a coward, too. I don't know what took place after that. I was in bed with Sherry, pretending to be asleep, when I heard Jim go into our room sometime later.

We had lived in their house for almost a year, and Carol and I never became friends. She was younger and sexy. Carol, Jim, and Jerry enjoyed socializing in bars and spent hours there. Jim was the only one that got noticeably drunk. Without Jerry, I don't know how Jim would have made it home sometimes.

Now the pieces fit together. Jerry's accusation explained why Jim came back north. His obsession about *my* having an affair justified his own unfaithfulness. I learned later that their affair started while I was in the hospital having Cathie. The first time Jerry confronted him about it, Jim denied it. That's why Jim was anxious to go to Florida. After he got me in Florida, he went back to Massachusetts. Had Carol sent for him or was it his idea? I don't know.

What didn't make sense was why he didn't leave me in Florida. Why bring *me* to live in *her* house while they were having an affair? Did he enjoy humiliating me as he had done when he brought both me and then Sally to Easter dinner?

I could not keep living here. Where would I go with four kids? I had a little savings in a Florida bank that Jim didn't know about. Was it enough to get an apartment? What about a babysitter? I'd need to get a different job, too. What I needed was to go someplace where he couldn't find me. Where? Would I have to ask Dad for help *again*?

If my TB came back, what would happen to my children? The thought of their being separated was too painful to consider. Only Jim could keep them together. If I could get past the death sentence, maybe I could beat the odds again.

I was sitting on the bed trying to decide what to do when Jim came into the room and told me that (miraculously?) the V.A. loan came through. "I just came from there. The people defaulted on the loan and moved away in the middle of the night leaving their furniture behind. All we have to buy is a television. I think we can find a secondhand one."

He offered no apology or explanation for the previous night or for his affair. It was as if nothing had happened. I suspected the V.A. house loan had been available for a long time. He preferred living in Carol's house where they were often alone. I had forgiven him about Sally, telling myself their divorce was over some argument and they were still in love. But there was no excuse for *starting* an affair with your best friend's wife. I would never trust him again.

I needed time to build up more savings. I never saw Carol or Jerry again. Neither of them was there when we went back to get our things. Among those things was the piece of paper George had given me long ago in the tuberculariium. I kept it, instinctively knowing I might need help someday.

I was not excited about having my own house. With Jim, nothing was permanent. I treated it like the rented places we lived in. Yet, I enjoyed being in a house where I didn't have to accommodate someone else's routine. As an unspoken apology, Jim bought me an electric washing machine. It wasn't as nice as Grandma's or Carol's but made my life a little easier.

I don't remember how I found Barbara. She was an older teenager who lived a few blocks away and became my babysitter. I'm not sure why Jim agreed to let me handle the babysitter, food, utilities, and transportation money from my paychecks and give him the remainder to "pay our other bills." They consisted of the house loan, his cigarettes, and his bar tab. If I needed anything else, I had to ask when he was sober or in the early stage of drinking. If I had a good reason and he had the money, he would give it to me. I asked only when it was necessary. I managed to save a few dollars each week that he didn't know about. I tried not to dip into that secret fund.

Our house was across from a park, which made a great place for the children to play. Under different circumstances, it would make an ideal home. Both Jim and I had new jobs and worked all that summer. I was working days in Quincy, so I only had to walk about two miles and take one bus. I worked in obstetrics, so one of my duties was to wash syringes and needles and wrap them for sterilizing. Often blood clots had to be removed from needles and the points sharpened. Nurses today never reuse syringes or needles, but that's how it was done throughout my nursing career.

Jim continued working through the summer and early fall doing landscaping. As winter approached, mostly tree-climbing jobs remained and he refused to do that. He was supposed to babysit while I worked, but he often called Barbara because he had to go somewhere. The nights he came home reasonably sober, we slept in the same bed. When he was really drunk, whoever brought him home dropped him on the floor inside the door or got him on the living room couch. He lay wherever he landed until he slept it off. I no longer cared.

Money was tight that Christmas. I had only a few dollars to buy one Christmas present for each child. Jim got a small tree on Christmas Eve when they were almost giving them away. He brought it home and then left to spend Christmas Eve with his bar buddies.

I was decorating the tree with strings of popcorn when my father arrived with toys for the kids. He had some way of keeping tabs on me because he knew about my black eye and I guess about our tight money situation. On the rare occasions we talked, it was by telephone. I know he wondered why I didn't leave Jim. People don't understand you need money and help to walk away with four small children.

When Jim came home, he saw the toys. "Have you been hiding money from me?"

"My father dropped them off a couple of hours ago."

"Still Daddy's little girl, huh? Why don't you grow up?" He went upstairs.

A few days later, I saw a folded paper on the floor by the chair where he threw his clothes. It must have fallen out of a pocket. It was a love note from Carol! They were still having an affair. I knew not to confront him when he was drunk. I must wait until the next sober opportunity.

As it turned out, I didn't have to confront him. He came home late that same night and said, "Carol's pregnant and it's my baby." He was bragging about it. "When the baby's born, I'm going to take it and we'll raise it."

"Absolutely not!" I said, remembering his threats to take Skip from me. "You can't take a baby from its mother, and if you could, I won't raise her child."

"Why not? I took your bastard and even adopted her. What's the difference?" With his anger rising, I said nothing more. He was too drunk to reason with. I had a pounding headache and had felt tired and nauseated all day. I just wanted to go to sleep.

The next day I felt worse. I thought it was stress over finding the note, but then I had been tired and nauseated before I found the note. Was TB coming back? The thought made my head throb. I comforted myself with the realization that TB had never made me *feel* sick. Could it be that the tubal ligation didn't work and I was pregnant again? I was scared. A pregnancy now could bring a death sentence.

Flu had passed its peak but was still going around. I prayed it was only that. I called the hospital to tell them I had a bad case of flu and couldn't work. Each day I became sicker until one day I felt too weak to go downstairs. I sent Sherry to get Barbara. She helped me downstairs to the living room couch where I lay unable to do anything except vomit any food or fluid. Barbara watched the children and fixed them lunch. Even the smell of food brought on dry heaves because there was nothing to vomit.

Jim must have been working because he brought home hamburgers or fried chicken each night for supper. I slept through a lot, unaware of what was going on. Then someone persuaded him this was something more serious than flu.

He found a doctor who agreed to come to our house that evening. Jim and the doctor carried me upstairs to bed. There was a lot of talking and bustling about to find wire coat hangers from which the doctor fashioned a way to give me intravenous fluids. I heard him tell Jim I was dying of dehydration as well as hepatitis.

Hepatitis B is spread through kissing, drinking from the same glass, using the same utensils, or from contaminated needles. It has an incubation period of 45 to 90 days so is difficult to trace. Today it is most commonly spread through oral sex or addicts sharing syringes and needles. People working in hospital labs and drawing blood were unknowingly caring for infected patients. I was forever pricking my fingers while cleaning and sterilizing needles. Jim, Barbara, and the kids had to be inoculated.

No local hospital would take me because hepatitis was highly contagious. The only hospital treating contagious diseases was in Boston. I barely remember the long trip in someone's car and nurses putting me in a wheelchair and lifting me onto a bed in a private room set up with the now-familiar isolation equipment.

I don't remember Jim leaving. No one visited me. It was a long way to the hospital, and we had no car. Mostly I remember the number of people who died from hepatitis or polio. When I heard "code blue" on the intercom, nurses hustled around. If it was on my wing and they closed my door, it meant they were removing a dead person. I knew the next death could be mine, and yet

I believed that somehow I would beat the odds again. I had no fear of dying. I had lived with that possibility too long, but I constantly worried about what would happen to my children. I was the only one they could truly depend on.

I don't remember praying for God to make me well. I had come to doubt God answered personal prayers. I decided he gave us brains and the ability to act so we would solve any problems we created. I forced myself to eat food that tasted like cardboard. I kept telling myself I was going to get well.

And I did. When I was released, Jim found someone to pick me up. I don't think the hospital bill was ever paid.

One day when I was feeling better, Barbara said, "We thought you were going to die. I asked Jim what would happen to the children, and he told me he knew someone he could move in with who owned her house. She would take care of the kids."

I knew he meant Carol, but I wondered if he asked her or if this was just his plan. I should have felt relieved that he had made plans for the kids when I died, but I didn't like his plan. I wanted to believe I had beaten TB. My great fear was that he would get a divorce and use my illnesses to get full custody of "his children." When he said that, I knew he was not including Sherry.

I had truly messed up my life and brought into the world four innocent beings that depended on me. The doctor said I could not go to work for two months, and for the first month I did little more than lie on the couch and make plans for leaving. I told no one of my plans.

We didn't have safe houses where women and children could hide. I was terrified Jim might beat me if he caught me leaving. He didn't have to hit me; fear of it was just as effective. When he was really drunk, he looked at me with that look I'd seen the day he did beat me. Drunk and sober, he warned me to never take his children away. Maybe he sensed my intention. Getting away took money. I was too ashamed of my bad choices to involve my father again. I needed to do this without his help.

I collected a week's paycheck from the hospital where I was working when we lived with Carol. It was still in my file. I never gave them an address when they hired me. Once paperwork was filed, I knew they probably would not discover it until we were living in the house Jim promised. Now, I gave them Barbara's address, and they sent a new check that I forwarded to my savings account in Florida.

I went back to work at a new job the first of April, sooner than I was supposed to because it was difficult pretending everything was fine. Jim was gone more. I don't know when he was working, drinking, or with Carol. I no longer cared. I didn't know if he was paying the house loan, and I didn't care. If there were late notices, he was getting them at another address. I set the last Friday in April as my escape date, knowing he would never suspect I'd leave six weeks before school was out, and that was the day he expected me to get a full month's pay.

I called George's number from a pay phone at the hospital so the call wouldn't show on our bill. It wasn't George's phone, but they gave me his number. When I asked for help, he said, "What took you so long?" He suggested I come to Lowell where he lived. Jim would never expect me to go there. "Where do you live, and when should I come?" he asked.

I gave directions and said it had to be the last Friday in April. If this plan failed, I didn't want George involved, so I said I would call that Friday if Jim wasn't home. He said he'd be ready.

For the first time in our marriage, I was lying. I told Jim this job paid by the month instead of the week, so he expected me to have money the end of April. In the meantime, I was cashing my checks and hiding the money in the bottom of a Kotex box. I had to have cash ready to give him if my plan fell apart.

I carried a few bags of clothes and toys to our winter change room that Jim never used. I took what I felt was most needed, including one bag of pots, dishes, and silverware that I thought he wouldn't miss.

When I picked up the last paycheck, I didn't tell anyone I was leaving. If my plan backfired, I'd need this job that Jim thought paid by the month. Neither the kids nor their school knew we were leaving.

When the kids got home, I told them they could keep their school clothes on and watch TV as a special treat. While they were eating an early supper, I ran to Barbara's house and asked to use their phone.

I called George. Jim had not come home, but in case he arrived before George got there, I said, "If the outside porch light is on, it means he is not here."

George promised to come as quickly as traffic allowed. Barbara said, "You're leaving? Good for you."

"I didn't tell you in case Jim questioned you. Don't come back. He doesn't know where you live. I know I owe you a month's pay, but I need it right now. I promise to send it as soon as I can."

"Don't worry," Barbara said. "Consider it a gift to the kids."

I thanked her, but it was an obligation that I would eventually pay.

It was almost three hours. George should be here. Couldn't he find the address? He had been away from this area for a long time. Was Jim still at the bar? Unable to resist the temptation, I called his favorite bar and asked if Jim was there. "Yeah, been here quite a while and he's getting drunk. Want to talk to him?"

"No! Please don't tell him I called. He'll be mad at me for checking." The bar tenders knew how quickly Jim could change from laughing to fighting. I didn't think he'd tell Jim, but if he did, chances are it wouldn't stop Jim from drinking now. If he remembered when he got home, I'd need an excuse for calling.

George drove up. I told the kids we were going with him, and they could each take one toy they could carry. While they were doing that, George and I loaded the bags I had prepared into the trunk. We turned the TV and porch light off. I had to leave so much behind. I also left a note on the kitchen table, saying: "I'm leaving and not coming back. Nobody knows where I am going."

I didn't know how hard he would try to find me, but I knew sooner or later he would try. By taking the kids out of school early and without notification, he couldn't track me through school records until I re-entered them in another school. That gave me four-and-a-half months to make permanent plans. George took me to his house for the rest of that night. This was the second time he was my savior when I needed someone.

The next morning, his wife watched my kids while I found a job and furnished apartment within walking distance of a hospital. The next afternoon, George helped me move into a small third-floor apartment next to a bar. Knowing how I hated drinking, it was the last place Jim would expect me to be. An elderly neighbor on the floor beneath me promised to watch over the kids while I worked. And Sherry was 10 now and very dependable.

I made them sandwiches before I went to work. I worked evening shift because the kids would not go to school here. I called Dad and told him I wanted to leave Massachusetts before school started but had not decided where to go.

He suggested I call Gladys' oldest daughter, Linda. She and her husband lived in northern New Mexico on a ranch where Stanley worked with the goal of buying the ranch. I'd met Linda, but we never had an opportunity to know each other. She was married and had a year-old daughter.

Linda agreed to babysit if I helped Stanley buy the ranch where he was working. It was near a small town with a hospital, and there was a small cottage on the ranch where I could live with my kids. It sounded ideal and far enough away that Jim might not try to find me. I applied for my New Mexico license and planned to be there when school started.

Before we left, I enrolled the kids in school in Lowell so their school records would go there before being forwarded to New Mexico. I gave them a fake address in Lowell to cause further delay when I was ready to enroll them in New Mexico. I hoped the records would be lost between schools. They could pass a grade entrance test if necessary.

I told George I was moving to New Mexico but didn't tell him where. I doubt Jim had any idea where George lived, but I didn't want George any more involved than he already was. I hoped someday he would contact his mother at the Florida address I gave him. I don't know if he ever did. His wife's family considered him their son. I never thanked him enough and lost contact. When you are on the run, you burn many bridges behind you.

We took a train to Albuquerque. Linda and Stanley met us as planned. Even before she spoke, I had an eerie feeling my life was taking another wrong turn.

"We've had a change of plans. The ranch deal fell through. I didn't tell you because I was afraid you wouldn't come. We can move in together, and you and Stanley can work and I'll babysit."

I wasn't keen on living with another family. Yet, I was already here, and I needed a babysitter. The next day, Stanley watched the kids while Linda and I searched for a house we could afford that would accept five children. We drove further and further outside the city limits before we found a furnished two-bedroom house with a fenced yard for the children. Stanley, Linda, and Rosa had one bedroom; my kids and I shared the other.

I paid the first month's rent and bought food and staples. The next day, while Linda, Stanley, and the kids moved in, I found a day job. It required a long walk to the bus stop. I was used to that.

I didn't understand why Stanley couldn't find work. He charmed people into believing white was black and claimed he was a jack-of-all-trades that could do anything. Linda and Stanley were arguing about it all the time. Nurse's pay was low in those days—not enough to support three adults and five children.

I was as frustrated with Stanley as Linda was, but at least working allowed me to escape their fighting. Linda took her frustration out on her baby, although she seemed fine with my kids. She spanked them with a hairbrush if they did something wrong. They weren't used to that but learned what they could and couldn't get away with.

Days became weeks and then months with our frustration mounting. Then Stanley announced he had been working on a secret invention. He actually got loans from two banks that he *convinced with nothing more than paper drawings* to lend him large sums of money to build it. They must have felt stupid when loan payments were never made and they discovered it only existed on paper.

I suspected there was no invention. He spent too much time at home. The invention never produced money, but the bank loans helped with the food for a while, and Stanley splurged at Christmas buying gifts for everyone.

My father, with limited education, always found work. I grew up with the misconception that *all* men were as responsible as he was. When Jim wasn't, I thought I had picked a loser because of excessive drinking. Now Stanley was even more irresponsible and he didn't drink at all. His lies sounded more real than the truth.

Stanley and Linda continued to fight, and the way she treated Rosa when she was feeding or bathing her was, in my belief, a form of child abuse. I think one reason Stanley stayed home so much was to take care of Rosa. When he was gone, Sherry often took over. Even at her age, she recognized what Linda was doing was wrong. I was proud of her.

At first we received letters; now there were phone calls from both banks demanding payments. I refused to make them. My paycheck barely kept up with the rent, utilities, and food. I was able to get a second part-time job at a nearby hospital, but half of each paycheck from there went into an escape bank account. Our living arrangement was on the brink of exploding. My children had heard Jim and me fighting, so I guess they thought it was something married people do. That was not how I was brought up, and it saddened me that it was all they knew.

By spring, Jim had traced me through school records to our post office box in Albuquerque. A town was all he needed. He called the hospitals, telling whoever answered that there was an accident and it was urgent to reach me. If it was the wrong hospital, they told him I didn't work there; when he found the right hospital, someone told him I didn't work that shift or put him through to the floor where I worked. Between school records and working in hospitals, it was easy to track me down.

He apologized about Carol, saying he loved and missed me and the kids and begged me to come back. He called so often my supervisor said, "Tell your husband to stop calling you at work."

His calls kept coming. He was in Florida living with Grammy, who reinforced he was working and not getting drunk. She never said he wasn't drinking.

Things between Linda and Stanley were at the boiling point. Most days when I got home, she was sitting in a chair staring into space with no housework done unless she told Sherry to wash the dishes. I had never known anyone in real depression and missed the signs. Instead of chastising her, I should have got medical help.

Something else was happening. I felt exhausted. As tired as I was, I didn't sleep well. I was angry for allowing myself to get caught in their problems. What started as someone to help *me* turned into three more people dependent on me. Worse still, in the back of my mind, exhaustion was a sign that TB was flaring in my lungs. If TB became active again, it would be resistant to treatment. That scared me.

When Jim offered to come and get us, I agreed. I knew it was a bad decision, but I had to get away from this situation. Jim seemed the lesser of two evils. I hated to admit it to myself, but in my heart there was still a spark of love. I don't know why. I thought about how he had treated me. Did I subconsciously want to be abused because I didn't feel worthy of a decent man? I was still hoping he could control his drinking and become the man I fell in love with.

Neither staying here nor going to Florida was a good answer, but I was leaving with secret escape savings.

Linda wanted Sherry to stay and help with Rosa. Sherry had friends in Albuquerque and bad memories of Florida schools, so I agreed. If this didn't work out, I'd be back.

Jim promised I could handle the money as long as he had what he needed for car expenses—insurance and gasoline, cigarettes, and "an occasional night out." He *seemed* sincere about wanting me back, but I wondered how long that would last.

Shortly after we left, the police came to arrest Stanley. Linda said, "He went crazy. He was scared to death of prison." Whether or not it was all an act, he convinced the police, lawyers, and court that he was insane. He was sent to a mental institution. Linda now had some kind of government help for the two children. She might not have got that had I still been there and working.

(Many months later, Stanley and a female inmate escaped from the mental hospital. They went to a small airport, and Stanley convinced someone to let him take a plane for an hour. He gave them false I.D. and probably a check that bounced. He had flown a couple of times and fancied himself a pilot. He invented amazingly believable stories. By the time the truth was discovered, it was too late. No one knows where he was going. He crashed the plane and both occupants died. Linda eventually got a settlement and didn't have to pay for a divorce.)

Jim rented a duplex apartment. Except for the usual battle with cockroaches, the first few months were good. He worked steady for a lawn company; I stayed home a full month for a much-needed rest. During that month, Jim didn't drink at all. I thought our marriage could be salvaged, although it was never like our early days.

When he started drinking again, it was on Friday night when he got his paycheck. By then I had a job with easy bus access. After he made new bar friends, his drinking spread to the whole weekend. We made it through the summer, but when that landscaping job ended, he never worked more than occasionally part-time.

Linda pleaded for me to come back now that Stanley was gone. We could live together and she would babysit. My dad had been to California and fell in love with the flowers and people. I decided to go to Albuquerque, see what Linda's situation really was and decide whether to stay there or take all four of my children to California.

While waiting for school to end, I was getting ready. Jim found two packed suitcases in the back of a closet. What made him look there? Instead of the fight I anticipated, he said, "You don't have to sneak away. Tell me where you want to go, and I'll drive you there."

Wow! I never anticipated that. He did no drinking on the way to Albuquerque, and promised that after he dropped us off, he would return to Florida and live with his mother. Whether he would truly have done that, I will never know. Fate stepped in when we were nearly to Albuquerque.

Skip was sick. We found a doctor on the outskirts of Albuquerque who diagnosed rheumatic fever and admitted him to a hospital in Albuquerque. Jim insisted he had a right to stay until he was sure Skip was okay. How could I deny that?

He helped me find a basement apartment in walking distance of a hospital where I got a job. The kids could walk to school. Instead of leaving after Skip was well, Jim stayed. He wasn't drinking then and had found landscaping work, so he said if I switched to evening shift, we could get by without a babysitter. I'd watch the kids during the day, and he would evenings. He promised to leave at the end of the summer, but his promises were made in melting ice cubes.

Sherry was still living with Linda and Rosa. (She had overheard conversations and knew Jim wasn't her real father, but I did not know that until years later.)

By the time school started, Jim was drinking daily. He had been laid off and never tried to find another job. Now, in addition to paying for his gasoline and cigarettes, I paid liquor bills. He ran up a bar tab, which was the first bill I had to pay every week. He thought he knew how much money I made, but I had a raise he didn't know about. He didn't know about my growing bank account. I knew someday I would run again.

Linda and Rosa went to live with Stanley's mother, so Sherry moved back in with me. One night she slipped out a window and disappeared. Linda and I searched for her at the homes of all the friends we knew. They either didn't speak English or claimed they didn't know. Pictures of her only brought head shaking. After questioning us, the police decided she was a runaway and stopped investigating.

I started working day shift when school started. Jim could never be depended on in the evening. He kept warning me about *his* children. His threats could be the ramblings of a drunk, but I knew what he was capable of when drunk. During the winter, his drinking increased. All promises were forgotten.

Near the end of February, in the middle of an ice storm, he took Cathie and headed to Florida. I called the police to try to stop them, explaining he was driving drunk with her in the car. They asked a lot of questions, but the outcome was that they would have to catch him driving. My word wasn't good enough. By now he was probably out of their jurisdiction, anyway.

Then the officer said, "You said you aren't divorced, and there has been no custody assigned? He has as much right to take his children as you do." That scared me.

He drove drunk over icy mountain roads that night, and it is a miracle he didn't kill them both. He made it to his mother's in Florida. After the most worrisome 10 days of my life, he brought Cathie back to Albuquerque. Did he do it to scare me? Did he bring her back because Grammy told him he should? Or did he come back because he wanted his boys, too? I don't know if he knew the police couldn't stop him, but I did.

His drinking increased. He was driving home drunk, and I don't understand why he never got caught. He came and went at all hours of the day and night. He often called the hospital to make sure I was really working when I said I was. A few times he came home early "to catch me with my lover." His renewed obsession made me suspect he was having another affair. I hoped she could keep him satisfied so he would leave me alone.

He talked constantly about taking his kids back to Florida. My getting away this time would need careful planning because his appearances were so erratic. I knew he could eventually find me, but letting him drive me here made it easy. I would not make that mistake again.

I saw him check suitcases, but I wasn't using them. The clothes I planned to take were in one dresser where it didn't look suspicious yet would allow me to gather them quickly. I obtained a California nurse's license and even had a job waiting in a city I picked at random.

A newspaper ad gave me a new plan he would never suspect and would be easier for me with children, as well as future working. I would learn to drive! I looked at ads until I found a car I could afford. In the same newspaper was a driving school ad that claimed if I took six lessons, they guaranteed my license. They were thinking of a teenager, not someone 34 who had never driven and had no mechanical ability.

Terry, my driving instructor, became my confidant. I would never have succeeded without his help. He took me to buy the car and register it. This took a chunk of savings, but lack of transportation had been my worst obstacle in the past. Terry made many compromises for me. He kept my car at his driving school; when I called, he picked me up at a gas station near my apartment. Neighbors never saw me get in a car.

Terry started with the basics. Unfortunately, I had no way to practice between lessons. Only Terry and my hospital supervisor knew about my plans because I needed their help. My supervisor had already guessed there was a problem.

Jim became increasingly verbally abusive, but so far he had not become physical. I knew each time he came home drunk that could change. Did he suspect I was planning something?

I was living on raw nerves and unable to sleep. Unless you have lived with an alcoholic, you cannot understand what it is like to constantly "live on the edge"—waiting for a call from the police station, waiting for a knock on the door, waiting for the door to open and a drunk looking for a fight to stumble in.

At my fifth driving lesson, I told Terry I needed to leave earlier than planned, even though it meant taking the kids out of school too early. He insisted I was not ready to take the driving test and needed more lessons. He offered to give them to me without charging more.

That same night, Jim came home earlier than usual but was very drunk. He called the kids together and said he was going to Florida. He asked who wanted to go. Nobody said anything.

"Well, you're going whether you like it or not." He turned to me. "I don't care if you come, but *my* children are coming with me." I thought he would forget it by morning, but he didn't. He got up just as I finished putting on my uniform.

"We're leaving today."

"We can't go today," I argued. "The kids have to finish school and I have to give a notice at work. We should wait for school to end. It's only a few more weeks."

"I don't need you. I'm going and taking my kids!" He slammed the door and drove away.

This didn't sound like his usual threats. He was sober and serious. I didn't dare go to work. I waited to see if he would come back after he thought I'd gone. When he didn't, I hoped he went to a bar and started drinking. There was the chance he was having a serious affair and might have gone to her house. Things were happening too fast. My window of opportunity could be closing. All I might have is now when the kids were in school and he thought I was at work.

I changed clothes and walked to the pay phone at the gas station to ask Terry to bring the car. While waiting for him, I called my supervisor, told her I was leaving and made her promise to tell Jim I was working in ICU and couldn't take phone calls if he called. She said my paycheck would be at the information desk whenever I got there.

On the way to the DMV, Terry gave me my sixth driving lesson. He insisted I wasn't ready. I knew he was right, but I needed that license. I told him I had to be gone before Jim got home. I had studied the code book for weeks, so passing the written test was easy. The lady marking it said, "Hey, you got them *all* right!"

The officer giving the driving test was not friendly in spite of my effort to engage him in conversation. Things started going wrong when he told me to back up between red cones on the parking lot. I could barely see them by stretching my neck, so I messed up. "Someone put them too close together," I said attempting humor. He didn't even smile.

Things went downhill from there. When he told me to turn right, the passenger wheel went up on the curb. Then he told me to parallel park between two cars. That's something Terry had just shown me how to do, but these cars were too close. He yelled, "Stop! You almost hit the front vehicle. We'll skip parallel parking."

I wondered if he sensed my relief. "Actually, I won't be doing any parallel parking," I said. "I don't mind walking a few blocks." He did not respond. (In all the years I have driven, I never again even attempted to parallel park.)

The officer barked, "Turn left at the intersection."

It was so quiet his voice startled me. I remembered Terry saying it is okay to turn even if the light is red. No traffic was coming, so I eased out preparing to turn left. I *saw* the car coming, and I know I had already hit my brakes before he yelled. As soon as I got through the light, he told me to pull over. I did a good job of that in spite of extreme nervousness.

"We're through," he said between tight lips. "I'll drive back." That was not a good sign. I stopped talking, and he said nothing until we were parked. "Who drove you here?"

"My instructor." I stumbled over the words.

"He'll drive you home. You need a lot more practice. You can try the test again in 30 days."

Somehow I kept my composure, but when I saw Terry my tears erupted. Between sobs I attacked this kind man as if it was his fault. "You promised me a license after six lessons. What am I going to do? I've got to get out of here today!" I think I was on the verge of hysteria.

"Sit here and relax," Terry said. "I'll be back in a few minutes."

I stared at the clock on the wall as precious minutes ticked off, one after the other. What was taking Terry so long? Had Jim already come back? If I didn't get the license, I was going to leave without it and hope I didn't get stopped. Time was slipping away, and I still had to get my final check.

The clock was reaching the hour mark when Terry came back and handed me my license. "There are stipulations," he said. "I promised him you would be out of New Mexico within 72 hours. Can you do that?" I nodded. "You must agree to trade this license for one in California as soon as you have an address. I gave them my word. Will you do that?"

"I promise I will." I hugged him right there in front of everyone.

He drove me to the hospital. While I picked up my check, he called his driving school for a ride. I never contacted him again. Another burned bridge.

I drove home, pulling into the area beside the house where I would be less visible. I was still loading things into the car when the kids came home. This time we took everything, including pillows, blankets, toys, and even one lamp we owned. Skip helped me load the bicycles and lamp on top of the car, tying them down as best we could with our clothesline. I guess we looked like homeless hillbillies.

We all climbed in, fitting ourselves between or on top of assorted luggage. I turned the key. The car wouldn't start. I had forgotten to check the fuel tank! Skip took money and ran to the gas station where Terry always picked me up. He had to buy a can to put the gallon of gas in. We poured it into the tank. Once more we got in. The car still wouldn't start.

While I tried to control my panic, Skip ran back for a second gallon. We poured it in. The car still wouldn't start.

I was in tears when Skip ran back for a third gallon. I think he was crying, too. The attendant said there must be something else wrong and brought Skip back in his pickup.

He got behind the wheel, turned the key, and magically, the car started.

"What did you do?" I asked.

He saw how upset I was and didn't make fun of me. "You can't start a car in reverse gear even if that's the way you want to go."

Then I remembered Terry said to always put the car in park before I start the motor. The gage was on the full mark. A third gallon would not have gone in the tank. During this trip I would learn how to drive.

Nobody said a word until I breathed a big sigh of relief when we were on Route 66 heading west. Even though Jim never expected I'd be in a car, I wanted to cover as many miles as possible before dark. I didn't leave a note. He would see I had taken everything. He probably thought we had gone with that lover he was obsessed about.

I couldn't waste money on motels. The kids slept in the car with the windows open; I threw a blanket on the ground on the passenger side.

That first night in the desert, under a million stars with no lights except from an occasional car traveling the highway, I looked up and whispered, "I beat the odds again, Peggy."

The TB doctor's threat, always in the back of my mind, had not come true, even though I certainly put it to the test. Maybe my tuberculosis was in permanent remission. Whether it was or not, I was going to follow Peggy's advice and get the most out of every day.

I realized I had made myself a willing victim for abuse. We teach people how to treat us. I thought I didn't deserve respect, so I didn't get it. Yet, inside, I knew that was not true. Yes, I had made bad choices. I had four wonderful kids that would never have been born if it were not for my mistakes.

My love for Jim was now totally gone. I wonder if it was always a one-sided love. I wonder if he was even capable of loving anyone or if it was sexual attraction and a need to fill the emptiness in himself that he never acknowledged.

Yet before alcohol took over his life, Jim was fun to be around and smart. Until the day he died, he had friends in the bar crowd willing to help him when he asked. Alcohol was a demon that possessed him until he died in his 40s.

I saw a falling star and didn't bother to make a wish. I no longer believed that making wishes or praying for God to fix my mistakes did any good. I promised myself that never again would I allow anyone to mistreat me.

Something I read somewhere flashed through my mind. The quote may not be exact: "A pessimist complains about the wind, while an optimist expects it to change. The realist begins adjusting the ship's sails." I was now a realist. I was still not the person I wanted to be, but I would no longer be a doormat for anyone. I had escaped not only abuse but also my own naïveté.

It was time to build a positive future for myself and my children. I was free to make my own decisions. The money I earned would no longer pay liquor bills. I would use it to have a better life for my children and for myself.

I had no idea what a wonderful life the rest would be.

CHAPTER 7

A New Beginning (1962 & 1963)

CROSSING THE BORDER into California was a disappointment. I had envisioned being welcomed by flowers and trees; there was only an unbearably hot desert (no air conditioning in the car). I already had my California license and a job waiting in Barstow. I managed to hold back my tears. Where was the California Dad raved about?

Skip was looking at the road map and pointed out that, if we went further south, a town named Long Beach was listed in my paperwork as having a veteran's hospital. It was on the coast near the water. Any place was more inviting than Barstow. I had worked at a veteran's hospital in Florida, so it made sense to keep going southwest.

Then suddenly my somewhat lonely highway became four lanes of bumper to bumper traffic with people switching lanes, cutting in front of me, and blowing horns to try to make me speed up. I inched my way over to the right lane where I saw there was a lane going off every now and then. I saw a motel sign at the next exit and pulled off and, shortly, into a driveway.

I was completely exhausted. I didn't have money for motels, but I realized the kids could no longer sleep in the car and me on the ground. We spent that night at an air-conditioned motel with a television to watch for the first time in several days. The desk clerk told us of a fast-food restaurant close by. Our habit was to eat one meal a day at a fast-food restaurant. We had Cokes, peanut butter and jelly sandwiches, and snacks the rest of the time.

I discovered Long Beach was a big city. Skip, studying the city map, was disappointed at how far we were from water. Then he found a coastal road that went through many towns that appeared to be at the edge of the beach.

The next morning we followed the coast, turning inland to check out a town. After that, we couldn't seem to get back to the coast. As we headed further south, we came to a sign saying Huntington Beach. It was a divided road. The island that ran through the middle had flowering bushes, plants, and trees, making it a strip of paradise.

I had found Dad's California. We marveled at the beauty around us until we came onto a short street lined with businesses that ended at the ocean. The kids were thrilled to see the ocean. A RENTALS sign caused me to pull over. When I inquired, everything was too expensive. Then the realtor said he had one very small house behind another house. It was furnished and affordable.

I said, "We'll take it."

"Don't you want to look at it first? It isn't in great shape."

He didn't know some of the places I had lived. I paid a month's rent, glad we didn't have to waste money on another motel.

It was one street over from the main street and had its own yard and parking place. There were two bedrooms, a living room, a small kitchen and one bathroom. The back bedroom didn't have wired lights, but Skip figured out a way to get lights. He was 11 years old!

"If we buy electric cords, a light bulb and an attachment, I can climb into the attic and run the cords from the front lights to the back ceiling that has a hole and attach a hanging light bulb."

I later learned his monkey-rigging wasn't safe—but it worked. This was our first of three California homes, all in Huntington Beach. The town turned out to be much bigger than we thought. We had only seen the old downtown tourist section, which had a small post office. I rented a box that gave me an address without disclosing our real one.

Next I applied for a California driving license. They accepted the New Mexico license and only asked me to take a written test on California rules. I missed the answers about driving and drinking because I had skimmed over them thinking they didn't apply to me. Still, I got enough answers right to pass. This time I would have passed the driving test, too, except parallel parking.

My first job was at a local hospital where I worked only long enough to get the money I needed to get established. My goal was to work at the veterans hospital in Long Beach before school started. The pay was better and it was further from Huntington Beach where the kids would go to school. I was still concerned about Jim tracing me, even though California was a long way from Florida. A divorce cost money and would help him find me, so I delayed getting it.

California pay was better, but the biggest difference was that no money went for liquor. It amazed me how much money I saved and still got my rotting teeth replaced with dentures. The kids had never been to a dentist, so I took them one by one to get whatever needed to be done. I was thankful none needed expensive work. I used a payment plan to establish my credit. There were no credit cards in those days.

When I transferred to the veterans hospital, it meant driving to Long Beach and back each day. I needed a car that had air conditioning. I knew nothing of the value of cars and had no one to help me, so I looked at the prices marked on the windshield with chalk until I found one I could afford that had air conditioning and automatic transmission. I made a down payment and arranged to make six payments. I never questioned the windshield price.

I was working night shift when I bought the car. At the V.A. hospital, we rotated shifts each month. I had taken the kids for a ride that afternoon, and everything was fine. On the way to Long Beach that evening, the car ran like a dream. On Sunday morning, I hardly got on the highway when the engine began making a noise, and the car tried to wander all over the street.

I remembered reading about someone whose engine wasn't bolted in tight and fell out while the driver was going down the street. I pulled over and went to the front and lifted the hood. The engine was still there and was no longer making strange sounds, even though I had left the motor running. I got back behind the wheel and started off again. Immediately, that noise was back and getting louder.

A few people on the sidewalk were looking at me and making gestures I didn't understand. I looked in my rearview mirror and, sure enough, there were parts of the engine flying from my car! I dared not stop for fear I wouldn't be able to start the motor again.

I headed straight to the dealer. The noise became increasingly louder and the car was now "shivering" violently. I was convinced the salesman had taken advantage of a woman alone to sell me a broken car.

I knew I'd guessed right when I pulled into the dealer's driveway because all the salesmen came outside and were laughing. Usually I can hold my temper, but now I was furious. I stopped the car beside them, jumped out, and slammed the door expecting the engine to hit the ground when I did.

"You should be ashamed selling me a car that has the engine coming apart just because I'm a woman."

They laughed even harder.

"I'm going to call the police!"

Then the salesman who sold me the car grabbed my arm and pulled me around to the passenger side. He was still laughing as he pointed at the back wheel.

Wow! There was no tire on that wheel! Just the round thing they mount the tire on, and it was squashed out of shape.

He took me inside, gave me a cup of coffee, assured me they would check to be sure the engine was bolted in and would not fall out. They *gave* me a tire with a center (rim) in the right shape. It didn't cost me a penny. They were so nice. I was glad we had chosen Huntington Beach for our new home.

I worked at the veterans hospital for the rest of that summer and the first school year. My plan was that, when school vacation came again, I would have enough money saved to take the kids on a six-week camping trip. When I returned, I'd get a job in another hospital closer to home. I hoped Jim would have given up trying to track me down by then.

I read about an organization called Parents Without Partners (PWP) open to all single parents. I had made no friends at work, and I no longer went to any church. I went to the meeting hoping I would make some friends.

When I entered the room, a few people smiled at me, and the president introduced himself and asked my name. "Have a seat. The meeting is about to begin," he said over his shoulder as he hurried to the podium. I sat in the back row.

He began the meeting by saying they had a visitor, announced my name, and said, "Let's welcome Kay." A few people turned their heads and smiled. Everyone clapped.

When the formal meeting ended, people immediately divided themselves into groups as they hurried to the finger-food table. I had not brought a food donation, so I remained seated. As I watched them, I wondered if everyone had forgotten I was there. A man and woman mumbled "Hello" as they passed me on the way to the door.

I guess that's it, I thought. They have all the friends they want. They don't need me. With a surge of disappointment that threatened to become embarrassing tears, the years of feeling like an outsider came flooding back. I started for the door.

One lady stopped me. Introducing herself as Betty, she began asking questions. Then she said, "I know it's hard to make friends when you first come out of a relationship. We've all been through it. I can help you break the ice if you come a half hour before the next meeting starts. I'll give you something that will help you."

If it had not been for one person taking time to offer friendship, I doubt I would have gone back. Sometimes all it takes is one person to start you on a life-changing path. At the next meeting, Betty met me at the door. She pinned a small plastic badge on my dress that said, "Official Greeter." Like something magical, it gave me the permission *I* needed to go up and talk to strangers. Why I needed permission, I don't know. Yet it worked, and I have been able to start a conversation with strangers ever since without a badge giving me permission.

I became an active member of PWP and for a year served on the chapter's board of directors. I continued working at the veterans hospital on the night shift all winter. It meant I was home in time to see the kids get off to school, and I was there when they got home. Since it was the least popular shift, I was not required to rotate shifts.

My only problem was difficulty sleeping in the daytime, so I often went back for a few more hours sleep after I fixed supper. The evenings of PWP meetings, dances, and discussion groups, I would forgo the extra sleep. This organization gave me the social life I craved even though I usually had to leave before the event ended to get to work by midnight when my shift started.

One night at a dance, I saw a tall, skinny, balding man standing by himself and looking lost. I used my official greeter badge to introduce myself and welcome him to the dance. His name was Joe Peterson, and he asked me to dance. We danced a couple of times, but there were never enough men at the dances, so I soon got cut out.

I was sitting at a table when a friend sat down beside me and said, "Kay, dump that man with the bald head. You can do better than that." She was about my age but dated only good-looking marines who were based in our area. She was still single the last I heard.

After a while, Joe came over and sat beside me. This was one of the few times the dance came on my night off, so I didn't have to leave early. Joe asked if I'd like to go for coffee afterwards, and I was delighted. He was funny and clever and his bald head didn't bother me. He was so tall that, even seated, it was mostly above my eye level.

After coffee, we sat in his car talking for a long time. I learned he had full custody of four children that were just a little younger than mine. His oldest boy (second child) and Cathie were just six months apart in age. I realized he was not like Jim or Stanley. He was a devoted father who would make a good husband.

He didn't seem to mind that I was raising three children, but when I told him about quitting my job and using my *savings* to go on a camping trip with my children, it turned him off. I didn't know he had been through a marriage with an irresponsible wife, and my vacation plans were a red flag to him. I did not tell him that all my bills were paid ahead or that I thought it was time to make another job change. And I didn't tell him that a couple of months camping could throw Jim off my trail as well as deepen our family bond.

The camping trip was wonderful. We had some car problems, but I handled them with the help of a few kind mechanics. We spent the most time at Zion National Park, camping with sleeping bags on the ground and a blanket between trees for privacy. Skip led the way hiking. Sometimes Cathie and I couldn't keep up with this energetic boy, but Scott did. We stopped at grocery stores to restock our food supplies and filled our canteens from the crystal-clear waterfalls.

The best thing about that two months was that it allowed the children to see me as a fun person—a real person and not someone who was always working, always tired, and mostly unhappy.

I met Jenny at a PWP meeting before the camping trip. When I learned she was an attorney, I asked her how to go about getting a divorce. She agreed to handle it. I was not asking for financial help, so I hoped Jim wouldn't try to fight for custody. While I was gone, Jenny tracked down his address. By the time we got back from our camping trip, she said the papers had been returned and she had them in her file until the court appearance, which was a few weeks away. I would have to wait a year after the court hearing before I could get remarried under California law. I was in no hurry. I had no prospects. All I wanted was legal custody of my children.

When the date arrived, I met Jenny at the courthouse. She handed the judge the signed divorce papers. He looked them over from first to last page. Then he flipped them back and forth comparing two pages. Looking over horn-rimmed glasses, he said to me, "What is your husband's full name?"

As soon as I told him, he almost threw the papers at my attorney. "Did you look at the signature? These are signed by someone else. You'll have to start over. This time you have to run a notice in all Florida and Massachusetts newspapers for *four* weeks stating you are filing for divorce and requesting the husband to contact you. If he does not contact you in 60 days, you can proceed with notification of your attempt to locate him. Include copies of all the newspaper notices when you file again."

Adding up time frames, I probably could not be legally divorced until late 1964. That wasn't a concern, but the added expense of newspaper ads was. I discovered I was Jenny's first case since passing the bar exam. I could no longer work at the VA hospital because I took my retirement savings to pay for our camping trip. I had no regrets; there were hospitals in all the surrounding towns looking for nurses.

Then I heard about a desperate nursing home administrator looking for a temporary fill-in while his supervisor had surgery. It was a line of nursing I'd never experienced. Working with the elderly in a convalescent hospital (nursing home) was frustrating at first. I was so busy learning my duties and the patients that I didn't have time to check patients who continuously called for help and were ignored by the aides. The first few days were so stressful, I was ready to quit. How could I work here for a whole month?

By the end of the first week, when I knew everyone's name and problems, as well as the daily routine, I had time to answer those pleas for help. Some only wanted to talk to someone; others needed reassurance their bills were paid and they wouldn't be kicked out; most were lonely and wondered if friends or family knew they were here. The ones that broke my heart were those who sat in wheelchairs by the front door all day waiting for visitors that never came—that perhaps were not even alive.

I saw many ways to improve the caregiving, but I was only a temporary replacement. When the RN returned, the administrator invited me to take over the evening shift, but I declined. It would eliminate the PWP discussion groups that I thoroughly enjoyed, and day shift worked better with the kids. Other nursing homes needed RNs.

I found one that was close to home. I became the supervisor and could put into practice things I didn't like. Then I discovered it wasn't the administrator who made the rules I objected to. Nursing homes were owned by big corporations run by a board of directors only interested in the bottom line.

I learned the aides in my new nursing home were only allowed to use *one* washcloth and *one* towel a week per patient. The last supervisor told them to use a corner of the soiled "draw sheet" to clean patients who were incontinent. I immediately changed the rule. They could use as many washcloths and towels as needed as long as they were not wasteful. This drastically increased our laundry count. When the Board discovered it two months later, they sent someone to investigate the administrator, who wasn't aware of my change in rules. She called me in to answer to the angry board member.

"If you were a patient, how would you like to wash your face with the same washcloth they cleaned your butt with half a dozen times a day?"

She blushed. I knew if she fired me, other places would hire me, so it empowered me to fight for what was right. With the administrator's help, we reached a compromise. Aides could use a second washcloth and towel for incontinent situations, but other patients still had one per week. It wasn't all I wanted, but it was a small victory. I loved working with elderly patients and decided the defenseless elderly would become my future nursing career. I lost more battles than I won because mine was a weak voice against big corporation policies.

One place I worked was a new type of combined assisted living and nursing home. I was appalled to learn they turned over their savings and all future social security benefits and other income. In return, the combination facility would take care of them for the rest of their lives. It sounded like security, but the patient was stuck there for life, regardless of changes in staff or whether a better personal option came later. I vowed if I were ever in a position to start such a place, the patients would not have to make a lifetime commitment. They could change their minds and leave at any time without losing any money.

I also worked at two general hospitals "on call" on my days off. Nurses' pay was low in those days, so even though jobs were easy to get, it was harder to make ends meet with two boys reaching their teens and wanting to participate in tennis and band. Both boys had become avid surfers, and that equipment was expensive.

I continued attending PWP dances and discussion groups. Joe Peterson attended many of the same discussions. I discovered we shared many beliefs and interests. However, with six women to every man, he was in great demand. He became part of the "in" group that had a party at someone's house every weekend. I was never invited until one time Joe invited me to a party at his house. It was a complete turn-off to us both.

A friend convinced me to bleach my hair because "blondes have more fun." I was only blonde one month, but it happened to be the month of his party. Seeing me with bleached hair disgusted him. My hair couldn't take it either and was falling out. I wasn't having any more fun as a blonde, so I went back to being a brunette and coloring the many grey hairs that were taking over.

Bleaching my hair and my irresponsible camping trip pushed me onto the friendship list but not a possible romantic interest. What turned me off about him were the weekly parties where everyone drank. I learned from Jim's family that when everyone is drinking except you, it makes you an outsider.

At the party at Joe's house, in an effort to fool people, I had Joe fix me a drink when I arrived. I carried it around all evening, swapping it with someone else's glass that had less liquid than mine. When I swapped glasses, the person didn't seem to notice. I thought I had fooled everyone, but Edna caught on. As I was leaving, she called me aside and confessed she didn't drink either, but hardly anyone knew it. She fixed her own drink using only the mixer without the booze and suggested I do the same.

However, it wasn't necessary. Nobody, including Joe, invited me to another party. Even though I would have found an excuse not to go, it showed me I was still an outsider. I remembered the first party I attended in high school. Now, I had that same feeling of not belonging.

I continued going to the monthly dances and all the discussion groups with little expectation of finding a marriage partner. I had a few close male friends, but none were possible future partners.

Joe's popularity helped his wounded ego, but that along with weekend partying eliminated him. Bob visited me a few times a week to ask my advice on his latest custody problem. I think he was still in love with his wife although he dated other women. He told me all about his latest girlfriend. Bill couldn't work because of back problems and was trying to get on disability. He was a nice guy and my kids liked him, but I wasn't getting involved with someone I could end up supporting. John was a recovering alcoholic and although he had been sober a long time, he was still one drink away from turning to alcohol again. Frosty did not want to raise more children. I think they all thought of me as their good friend, but none of them ever asked me for a date. Perhaps I was sending a subconscious signal that none of them was the partner I was looking for.

I settled for their friendship.

My luck with men was dismal. I was better off single. I had beaten TB, I could support my children, and PWP gave me a social life in which I was accepted as a friend to both men and women. With working one full-time job that I loved and being on call at a hospital on my days off, I had a busy and happy life. I slept well without fear of a drunken husband waking me.

CHAPTER 8

Joe's Story (1927—1963)

CARLSBAD CAVERNS NATIONAL Park, located in the southeastern corner of New Mexico, had been there many centuries before Jim White, a cowboy, was credited with discovering it in 1901. The high desert land was populated with lots of mesquite and other spiny desert plants. The only trees were scrub cedar, which stood scarcely head high. Yet under this harsh desert mask there was a fairyland called Carlsbad Caverns.

It was formed over two million years ago by water entering fractures in the earth's surface that gradually hollowed out the caverns into a network of tunnels. As rain filtered down through the acid soil, seeking cracks in the limestone, drops of water gathering in the cavern ceiling left tiny rings of limestone as they evaporated. With the passage of centuries, these limestone rings grew into icicle-like stalactites, while the drops that fell to the floor grew into stalagmites. Here and there, nature patiently joined them into wax-like pillars. In 1923, President Coolidge declared Carlsbad Caverns a national monument.

To reach the national park, one must climb a narrow seven-mile winding road through the limestone hills where even hardy desert shrubs spread their roots in a death hold on the canyon walls. Behind what is today a beautiful, modern visitor's center are modern homes where park employees live.

It was very different in April 1927, when Thomas Peterson brought his young wife, Lena, and eight-day-old Joseph to this national park where he worked as a laborer. In those days, the houses for higher-paid employees and rangers were stone; the other employees lived in tarpaper shacks, which they usually built themselves. Thomas, Lena, Joe, and his eight-year-old sister lived in the shack Thomas built down by the bat cave. He wanted to be away from everyone else.

Neither Thomas nor his ancestors had any formal education. Their family was uneducated laborers in Bristol, England, and expected to be the same when they migrated to the U.S. They became sharecroppers who lived and worked on farms belonging to other people. Thomas and his two brothers grew up working in the fields instead of going to school. Of the three boys, only Thomas taught himself how to write his own name and could read a little.

Perhaps that is why he was able to get a steady job as a goat herder on a farm in the hill country of Texas where Lena's family lived. Her mother died when she was 13 and she became "wild." She was 17 years younger than Thomas, educated, and from a respectable family in the community. Although Thomas had little to offer a socially-minded young lady, they married and moved to New Mexico where Thomas was able to get a laborer job at the newly acquired national park.

Thomas was a dependable worker who never asked for more money and was never given more in spite of his hard work. He felt lucky to have steady employment in the Depression.

Joe's only sibling was his sister. It became her job to care for her brother while their father worked long hours building trails on the cavern floor and their mother entertained herself, visiting neighbors or drinking at the local tavern. Like Jim, Lena loved the barroom social life. She seldom had money herself, but she was fun, so her drinking buddies often paid her tab. She was often invited to go on a trip that could last for weeks.

Joe didn't think of her as his mother. They were two people living in the same house. He doesn't remember ever calling her anything but Lena.

By the time Joe was a toddler, his sister was tired of taking care of her little brother. She let him do as he pleased. He got himself into and out of trouble without his family's knowledge or help.

Joe's earliest memories include climbing rickety steps down into the bat cave whenever the west wind blew hard. Both children feared it would blow their house down. The bat cave was always cool, so it was a great place to play on a hot summer day.

By the time Joe was born in 1927, people from all over the world were coming to New Mexico to see this wonder of nature. Joe said, "As I got older, I hung around the visitors. I liked talking to tourists because I was hungry for knowledge of the world. (No televisions.) I told them all I knew about the caverns hoping they would reciprocate by telling me about where they lived. Then I'd go home and try to find the places on a map."

Often someone would say he was lucky to live in such a famous place.

They didn't know how lonely it was with only a few children living there. He entertained himself with whatever was available. He was sure-footed and began climbing and jumping over cactus. All that was available in the desert on top of the caverns was cactus, snakes, and tarantulas. And, of course, rocks. Lots and lots of rocks.

One day, Joe spied a skunk foraging in scrub cedar. A target like that is an invitation for a five-year-old boy. He was having a good time practicing rock-throwing when the skunk turned around and sprayed like a hose under pressure. The skunk's aim was better than Joe's; the game ended in sudden victory for the skunk.

Unable to stand his own stinking body, Joe raced away in search of a shower. They had no shower at his house and "washing up" in a bucket of cold water wasn't going to get it. He made what seemed a logical decision. All the rangers were working, so he used their shower room.

He showered with his clothes on using their soap and shampoo until all the hot water ran out. He felt much better. If he stayed outside in the fresh air until dinner time, maybe any lingering smell would be gone and his parents would never know. He had not thought about what would happen when the rangers got off work.

It became one of those "Oh, no!" experiences punctuated with very loud, very angry, and very descriptive language from the rangers whose shower he used. Joe took their threats of "We'll kill that kid!" seriously and stayed well away from all the rangers for several weeks.

"As I became aware of what little I knew about our family history, I realized my parents should never have married. It wasn't just the age difference. Dad was a very quiet man who seldom said anything but listened carefully to what other people said, while Lena chattered away and never listened to anybody. Her moods were unpredictable. Sometimes I could get away with anything, and other times she whipped me for no reason."

From the beginning, school was a problem for Joe and had an effect on the rest of his life. Because he got his growth early, he was always the tallest in his class during grade school. This made people think he was older than he was and led to the viewpoint that he was either mentally retarded or stupid.

He had poor eyesight, so for years he lived in a blurry world. When he was 12, his eyes were tested for the first time and he started wearing glasses. He was astounded when he walked outside and saw that big glob of green he called a tree actually had individual leaves.

Along with poor eyesight, he may have had dyslexia, which affected his reading ability as did ADS (attention deficit syndrome). Neither problem had been identified by the medical world.

Joe was curious about everything, and when he read something he would stop and think about the implications that went with it. While he was still digesting everything about the first paragraph in a book, the class had turned to the next page.

Thomas was unable to help Joe with even the simplest school work. His mother could but wouldn't because "teaching children was what they paid a teacher to do," and she had something more important to do. Joe turned to his sister for help. Sometimes she was patient and helped him, but usually she was involved in her own homework and friends. With no one to care whether he had any homework, it was more fun to play after school.

Schools were different in those days. Teachers were allowed to use whatever punishment they deemed necessary. Consequently, children paid attention, didn't fall asleep at their desks, or make rude remarks to the teacher or other classmates inside the schoolhouse.

They also often didn't have enough books, so two kids shared a book. Joe was in the second grade when he and Elizabeth were reading a story about a squirrel burying nuts for the winter. They started trying to find words that rhymed with nuts. Joe came up with the word "guts."

Elizabeth ran to the teacher saying Joe said a bad word. She whispered the word in the teacher's ear. The teacher said, "Joseph go to the cloakroom." That is where punishment was doled out, and Joe was scared. By the time the teacher got there and asked him what bad word he had said, he couldn't remember. Actually, he didn't know guts was considered a bad word.

The teacher knocked his head against the wall several times, demanding he tell her the word. He tried with all his might but simply couldn't remember.

Thinking he was just being defiant, she banged his head every time he said, "I don't know." This would go on for a few minutes, and then she'd go and teach class. When she came back asking the same question, the head banging would begin again. Finally, the teacher gave in and said, "You said *'guts'* and you better never say it again."

By now the back of Joe's head was full of knots. He promised himself he'd never forget that word again; he kept saying it over and over in his head. He never told his parents about the incident for fear of further punishment.

He fell further and further behind, first in that week's lesson and eventually a whole grade behind. Nevertheless, his teachers kept passing him from one grade to the next because they knew he was trying. As he got further and further behind each year, he wondered if he could ever catch up. This made him *feel* stupid although he retained all he learned, whereas most of his friends did not. Fearing people would laugh *at* him, he made jokes about himself so they were laughing *with* him. He soon became the "class clown."

This made him popular. He taught himself to see everything in a humorous way. As the years passed, he built on this reputation, making him fun to be around. As for the teachers, most knew about his mother and felt sorry for this likable kid. When a teacher was talking, Joe paid close attention, and those were the things he learned—not what was in books.

If it were not for some friends, he would have had a miserable childhood. Two friends in particular brought him to their homes for a meal and sleep-over. Both families taught him some manners, and their food was always good.

"Sometimes I would lie in a friend's bed and fantasize how wonderful it would be if I had a family and a mother like my friend had."

Joe did some foolish things because he thought he could do anything, and no one pointed out the dangers. Among the memories of his childhood, one stands out because it nearly ended his life.

He and his friends did lots of hiking and climbing the foothills of the Guadalupe mountain range. Joe was the most agile of them all. One day just after he turned 12, he and Howard were climbing the smaller cliffs when they encountered a tall cliff that looked like a mountain from where they stood. Joe challenged Howard to climb up that cliff; Howard wanted nothing to do with such a dangerous climb.

Joe said confidently, "Well, I'm not afraid to climb it! You go around and I'll wait for you at the top."

Howard tried to talk Joe out of it, but Joe was sure he could do it. Howard watched until Joe got quite far up the face of the cliff, and then he started his longer trip around one side of the cliff to where it was easy access to the top.

The further up Joe climbed, the more obvious it became that this was not a good plan. Handholds became fewer and fewer; Joe's upward progress became slower and slower. He was scared. He knew the odds of climbing were against him, but he had to keep going because it would be impossible to climb back down. After what seemed an eternity, he heard Howard's voice calling him from above.

"Help! I'm in trouble," Joe yelled. "I can't find any more handholds!"

Howard climbed down to the top of the cliff and began telling Joe where to reach for handholds that he could see from the top but weren't visible to Joe. At long last, with Howard's help, Joe made it to the top. He was so tired and scared that he was shaking violently. He sat there for some time until he finally stopped shaking. "I would have cried," he said, "but how could I cry like a little girl in front of Howard?"

This was only the *first* time he escaped from something that could have killed him.

Joe was 13, with Europe already at war, when the government established a new rule that only U.S. citizens could work for the U.S. Government. Thomas was just a toddler when his family migrated to Texas, and neither he nor his parents ever saw a reason to become citizens. (Their inability to read and write would probably have kept them from becoming citizens anyway.) Thomas was given a notice of termination. He must move off the national park property.

Thomas couldn't believe the government would fire such a hard and willing worker. He believed the paperwork would get straightened out, so he continued working every day even though they said he wouldn't get paid. After two more months, he was "run off." He was confused and unhappy. This was the first time in his life he had been fired, and he didn't understand why.

His first job in town (Carlsbad) was at a tourist motel mowing grass and maintaining the shrubbery. The tightwad he worked for insisted he paint the steel that supported a very tall neon sign. Thomas was 65 years old and not physically able to climb that steel. Again he was fired and he was so ashamed, his mental attitude was never the same.

Next, he worked at the city creamery, washing milk bottles and scrubbing the concrete decks. In those days, milk was delivered to individual homes in glass bottles; people set their empty bottles near the door for the delivery man to pick up.

His sister had been living in town with a girlfriend. She met a handsome young soldier stationed there. They fell in love, were married, and she moved away. Joe also spent months at a time living with a friend's family and sleeping in their attic.

When Thomas moved into Carlsbad, he had to find a place to live. He found a row of old shacks that stretched back to the railroad track. These single-family houses had basically been thrown together, with each house in the row worse than the one before it. Thomas got the last of the row which backed up to the railroad track. With both kids gone and Lena away so much, Thomas often returned to an empty house at night.

Living in Carlsbad made it easier for Joe to get work. Like many boys in those days, his first job was delivering newspapers and telegrams and setting pins in a bowling alley. Joe made enough to pay his dad's electric bill every month, but he also saved his pennies with one specific goal in mind.

Joe was seven when his family took him to visit relatives who had a boy his age. This cousin had many toys, including an erecter set. Joe was fascinated by all the things they could build with it. How wonderful it would be to have something like that to play with.

In Carlsbad, erector sets were on display in the store window only at Christmas. He would walk by the store dozens of times, looking at the erector set and wishing he could have one. At age four, he had learned there was no Santa Claus, so there was no hope there. Each year, he begged his parents for that gift but was always disappointed to see a new pair of britches or shoes instead.

Now he could save some of the money he earned. It took a long time to save enough, but he stayed with his goal, dreaming about this wondrous thing he would someday own. He was almost 14 before he had saved enough. That Christmas, when the stores sold them again, he bought the set with the most pieces that even had an electric motor.

He took it home and excitedly opened the box on his bed. Spread out before him were all those shiny blue, yellow, and red pieces of different sizes and the screws to build them into wonderful designs. As he gazed at it, he realized it was a child's toy and no longer fascinating. He packed it away carefully in the original box and put it under his bed where it remained until his mother took it to a pawn shop for money to buy beer. That lesson remained with him all his life: "There is a time for everything, and if you don't do it when the time is right, it will be too late."

At 14 he worked at a gas station and later delivered water to the workers on a construction job. When summer ended, he started his freshman year even though he was not at that grade level. He tried to keep up with his class but failed all the tests. He skipped school sometimes to work but kept trying to catch up with his friends.

It was football season and Carlsbad High was playing their biggest rival, El Paso High. He and three other boys decided to go to El Paso and watch the game. They were all about 14 and rented a room in a cheap hotel. They paid an old boozer on the street to buy them a big jug of cheap booze. After the game, they spent the night in their hotel room, drinking and reliving the night's events.

The next day they went across the border to Juarez, Mexico. After roaming around town for a while, they decided to go into a bar. In Juarez, anyone could sit on a bar stool and order a drink as long as he could pay for it. And he could drink it at the bar just like grown-ups do. This really impressed the boys. Joe didn't remember how many drinks he had, but he was sure it was too many. The last thing he remembered was laughing and joking with the guys in a bar.

He woke up later in a small concrete room with bars on the front of it. He pushed on the bars and surprisingly they swung open. There was a wide aisle with identical enclosures on each side. At first he didn't realize he was in jail and started walking down the wide aisle until he came to steel bars that blocked his way. He climbed up the bars, intending to jump down from the wall and get out of there. When he got up there, he could see that there were more enclosures, so he proceeded to walk along the top of the two-foot-wide adobe wall. Suddenly, three shots rang out.

He jumped off the wall on the opposite side of where he thought the shots came from. "I was scared and didn't know what to do except hide. I tried several cell doors until I found one that was not locked. I went all the way back and hid in the darkest corner, listening to the sound of feet running by the cell and hollering at each other in Spanish."

Then his cell door opened, and in came a man who looked around but did not see Joe hiding in the dark corner. After more yelling and running, another man came in. This one saw Joe cowering in the corner. He went out, closed the gate, and leaned against it while hollering in Spanish.

Several angry guards descended on a scared American whose height made him look older than he was. They dragged him into the aisle and started beating the daylights out of the terrified boy. When their anger dissipated somewhat, they dragged him down the corridor to a huge cell, pushed him inside, and locked the door.

"The place was crowded with Mexicans. It was winter and very cold in the cell. There were two or three occupied cots in the room, but everyone else had a pallet of old blankets or anything they could get that would keep them warm."

He found an empty spot next to a wall and hunkered down. He was sitting on the floor shivering violently when a man on a pallet near him threw the covers back and indicated Joe could get warm.

"I no sooner got under the covers than I felt a hand trying to unbuckle my pants. I decided I wasn't that cold. I hastily crawled out and went back to the cold concrete wall. I shivered and dozed on and off the rest of the night."

The next morning a guard called, "Peterson." When Joe answered, the guard said in English, "Come with me." He led Joe to a room where an officer sitting at a table called someone to come forward. After the man pleaded his case, the officer apparently decided what action to take.

"It was interesting to watch even though it was all in Spanish. I understood enough to figure out what was happening."

Finally the officer called "Peterson." When Joe approached, the officer said in English, "You're charged with being drunk and disorderly. How do you plead?"

"I guess guilty. What will the fine be?"

The officer replied, "Twenty pesos or two dollars American."

Joe said, "I had money in my billfold, but somebody has taken it. Can I use a phone to call someone in El Paso?"

The officer said, "After we finish here, you can use my phone."

When all the other prisoners were gone, the officer told Joe he could use the phone.

"I called the hotel where I had stayed with the other boys."

The desk clerk said, "They checked out without leaving a forwarding address. But you had a call from a Marian. She left a phone number."

Marian was one of his classmates. He called the number. She heard about his being arrested and worried about what might happen in a Mexican jail. She was staying in El Paso with her aunt over the weekend. Her aunt agreed to drive to Juarez and pay Joe's fine.

"Marian and her aunt took me back to the hotel where I had stayed. My friends had checked out, but the room had not been cleaned, so I took a shower and cleaned up the best I could. When I examined myself in the mirror I saw there were bruises all over my body. I was ashamed of the experience and hoped no one else would find out."

However, the boys he was with had already told people. When Joe went back to school, everyone was talking about it. It seemed he had become a hero. Before long, Joe began bragging about it, too. "Whenever I mentioned it, people wanted details. I never went back to Juarez, and on the very few trips I've made across our southern border, I never got drunk. I beat the odds once because of a kind friend. I wasn't about to try again."

Once people stopped talking about his adventure, Joe realized he was so far behind his classmates he'd never catch up. He was failing everything but physical education. "I was wasting time. I dropped out of school and went to work as a laborer at a local potash mine. After a while, I was promoted and transferred to the electrical department."

He had paid the utilities bills for a long time; now he was making more money than his father, so they swapped positions. Joe paid the rent even when he stayed in town with friends for weeks at a time.

It was tempting to continue earning money, but Joe felt the need to do his part, so he joined the navy when he was 17. The navy sounded more glamorous than the army, but Joe would have made a good soldier with his ability to walk and run long distances.

He was sent to Southern California for basic training. While in training, his legs broke out with large purple spots. He was sent to the infirmary and a week or two later was transferred to the San Diego Naval Hospital where he stayed for weeks while the doctors attempted to diagnose his ailment. They came up with several possibilities, including a form of rheumatic fever, but each one had a question mark after it.

From the hospital, he was transferred to Rancho Santa Fe Recuperation Center, which was part of the hospital. "It was the best duty any sailor could get. The food was great—steaks, ice cream, and all types of goodies. When the purple spots went away, I was sent back to finish basic training. The doctors never agreed on a diagnosis."

After basic training, he was stationed on the *Lexington* aircraft carrier, which operated out of Pearl Harbor. The ship returned to Pearl Harbor every two to four weeks, which meant they could go ashore and spend their money on liquor, prostitutes, and tattoos. By now the war was winding down. After he was assigned to the *Lexington*, it never left the general Pearl Harbor area. He regretted missing any action and, because of that, didn't take advantage of many veterans' benefits. Yet, the time he spent on the *Lexington* was a life-changing experience.

There was a small library onboard. Joe had never been in a library and was amazed to find books on a variety of subjects. Once he picked up the first book that he could read at his own pace, he became addicted to reading. Someone else might read the same book in half the time, but Joe remembered details those others didn't.

It was in that ship's library that his true education began. He added to his knowledge for the rest of his life. He seldom read novels, even those based on history, because he thought it was too difficult to distinguish fact from fiction. He loved history books, especially biographies as well as geography and science, and always kept abreast of current news the rest of his life.

When the war ended, they discharged everyone who didn't want to continue a navy career. He and a shipmate decided to go to Alaska where "there was more opportunity to get ahead." Alaska was not a state then; it was a territory. Joe was anxious to start making the million dollars Alaska advertisers promised. Neither boy was 21.

They bought tickets in steerage class to Juneau. Since Juneau was the capital, they assumed it was the biggest town in Alaska. Actually, it was a small village.

They went to work at the Juneau spruce mill and found a place to stay at a rooming house. After two weeks, they realized they were barely breaking even, and a million dollars looked further away than ever. Both were depressed about staying in this expensive and boring town. They walked to the Alaska Steamship dock only to find they were on strike. On the way back to their boarding house, they passed an airline office. On the spur of the moment, they went in and asked when the next plane went out.

The clerk asked, "Where to?"

Joe repeated, "Out of here."

"There's a plane for Anchorage leaving in two hours."

"How much?" asked Joe. The clerk gave a price. His friend nodded. "Give us two tickets."

In Anchorage, they found work as construction laborers on a four-story administration building being built on the base at Fort Richardson. The wages were not great, but the food was, and they had clean huts to sleep in. It worried Joe that he was not learning anything. He worried about spending his life as a laborer.

Winter comes early in Alaska. The foreman told them they were closing down. However, those who wanted to continue working would be transferred to a company that worked through the winter. Joe accepted the offer. His friend went home.

The new company had better quarters with two men to a room. After a month, Joe was promoted to stove mechanic where he learned to repair and install oil heaters. It paid more than a laborer but was boring. He couldn't see spending his life that way.

He talked to the electricians' superintendent about being an electrician apprentice. He said he had some experience as an electrician apprentice in a potash mine in Carlsbad. He knew he made a good impression, but they were not hiring. Although disappointed, Joe believed persistence was important. Every couple of weeks, he went back to ask if they needed anyone yet and each time was disappointed.

Then one day an electrician came to the stove shack and told Joe the superintendent wanted him to come over. By coincidence, it was the same day the stove crew got word their department was going to be incorporated with the sheet metal shop. That meant stove mechanics were going to get journeyman sheet metal wages, which was the same as a *journeyman* electrician's wage.

Joe's friends believed he would turn down the apprentice offer because he would make less money as an apprentice than if he stayed in the stove shop. It was a difficult decision. More money now—or have a chance to learn a better trade. Joe told his friends, "I've learned all I can here. I want to get in a real trade." He started the next morning as an apprentice electrician and worked in Alaska for the next two years.

While working at the stove shop, he became friends with Buddy. Buddy was a child when his family became victims of the drought in Wisconsin and several other states. Many farmers in those states lost their farms, their houses, and even members of their families during a devastating drought that lasted for 10 years. One family who lost everything was the Lentz family.

In February 1935, President Franklin D. Roosevelt set aside 26,000 acres in the Matanuska Valley in Alaska for special homesteading for farmers on Federal Emergency Relief who were willing to move to Alaska and start a colony there. It was called the Matanuska Colony Project and, like the CCC, was a program Roosevelt started to end the Depression.

Roosevelt believed in a hand up, not a handout. They were given free transportation for the entire family. On arrival, they drew lots for 40 acres of land, a house to live in "that had all the necessary modern conveniences," and whatever tools, equipment, and livestock they needed to get started. In return, they had to stay there until they paid the government back $3,500. Payment terms were easy and could extend over a period of 30 years. The Lentz family was among the original 202 settlers and one of the few families who remained in Alaska for the rest of their lives.

Summer days have many hours of sunlight, yet the CCC builders were slow building the permanent houses. The colonists finally got permission to build their own houses; by helping each other, all the outsides were finished by winter. Whatever their former background or philosophy, they now had an established policy of helping each other and anyone who needed it. It was not surprising that when Buddy brought Joe home over a weekend, the family adopted him as if he were one of their sons.

Joe said, "They became my new model of what a family should be. I wished they were my real family. It would have been wonderful to grow up with parents like Ma and Pa."

Buddy had an older brother, Donny, and a pretty blonde sister, Joann. Joe and Joann fell in love. Joe thought they were headed to marriage until an older and more prosperous fellow began courting her. It was evident she was more interested in him. Joe and Joann remained friends, but not friends enough for her to invite Joe to her wedding.

While his competitor was courting "his girl," Joe was working in Anchorage with Buddy and Donny and returning to the homestead on weekends, where they attended the dances in Palmer. The three young men were popular because they were hometown boys who would be there when the military boys went home.

The family had a neighbor who owned a pig farm. Mr. H. told Donny that neither he nor his wife had ever taken a vacation because they had no one to look after the farm. Donny agreed to run it for two weeks and talked Joe into helping him.

They were going to run it for a couple of days and learn the ropes before the pig farmer left. They moved into a shack very near the pigs and settled down. Shortly before bedtime, Joe decided they needed a pail of water from the well and offered to get it.

"When I got near the well house, I heard a terrible noise coming from inside. I thought there was a bear in there. I had a Coleman lantern, so I carefully approached. The well-house door, which should have been closed, was wide open. I looked inside and found the well house empty. Yet something was making a terrible noise. Maybe it was inside the well. I tipped my lantern up to see down into the well. In the bottom was the biggest, maddest pig I ever saw."

Joe rushed back and got Donny, who thought Joe was playing a trick. When Joe kept insisting, Donny reluctantly went with him. Before they even got close, they heard the pig. Donny sent Joe to get Mr. H. and his grown son. The four of them tried to get that big male pig out of the well. Fortunately, the depth of the water was not over the pig's head, and even more fortunate he went in rear end first or would already have drowned.

Mr. H. said, "The only way to get him out is to get a loop of rope over his head and behind his front legs, so the rope won't choke him." After many attempts, they realized it could not be done from the top of the well. Mr. H. said, "Looks like one of us will have to go down to get the rope around him. We'll lower someone down in the well, low enough so he is over the top of the pig but out of the pig's reach. That pig is mad and has sharp teeth."

They looked at each other. Since Donny was the smallest, he was nominated. After hours of work, Donny was finally able to get a loop around the pig's head and one front leg. After they hauled Donny up, they tied the rope to a tractor and started pulling out the fighting pig. When they got him near the top of the well, they stopped to redo the ropes. After several attempts, they got a lasso around each of the pig's back legs and all pulled together until the pig was finally out. The farmer looked him over and decided he didn't appear injured.

"When we got Donny out of that well, he was a lot happier than the pig," Joe said and even Donny laughed.

Although it was Buddy who introduced Joe to the family, it was Donny and Joe who became lifetime friends. Joe spent the next year in Alaska helping Donny "prove up" on his homestead. A winter in a primitive cabin in Alaska was not like the comfortable Lentz homestead Joe was used to, but it was another adventure. Years later, when Donny moved to Washington to raise a family, Joe and Donny remained in touch and visited each other through the years.

Joe had been in Alaska a little over two years and wasn't making much headway on those million dollars he dreamed of. With Joann happily married, he decided to go home and finish his last two years of electrician apprenticeship.

However, the chance to make lots of money interrupted his apprenticeship again. His working buddies, Bob and Johnny, heard there was an overtime job in West Virginia. This opportunity to fill their pockets with overtime wages enticed them to quit their jobs and head east.

The day before they arrived, the job went on strike. Cursing their luck, they spotted a Bar and Grill sign. It seemed like a good place to drown their sorrows while they decided whether to wait out the strike or go home.

The bar was run by three young women. Over several drinks, they told their sad story. The lady owner asked if they wanted to help run the bar. After a few more drinks, it was decided that, with their help, the bar could be kept open 24 hours. Bob and the owner took the day shift; Johnny and the second girl would work the evening shift, and Joe and the owner's little sister took the night shift.

Since the boys also needed a place to sleep, the girls invited them to share their rooms. Joe said it was a good deal because his girl turned out to be a nymphomaniac. He really liked this girl and even considered marrying her. In the time they'd been there, the bar business kept declining, and they weren't making any money. This was not the kind of life Joe wanted.

When another friend stopped by on his way to a job in Ohio, Joe accepted a ride. After two days of thinking about it, he decided to return to Carlsbad and finish his apprenticeship. He wanted a permanent way to make a living. An electrician might never make a million dollars, but he enjoyed the variety of work, and it was a lasting trade.

Joe was flat broke when he told his friend he was going to hitchhike to Carlsbad. His friend gave him $20 for food because there was no telling how long it would take him to get home. After looking at a map, Joe decided to cut across on country roads because it looked shorter. It was a mistake. Rides were few and far between and were only for a short distance. His last ride ended just before dark.

He had not walked far on a deserted road when it started raining. He was cold, so when he saw a culvert, he crawled into it thinking he'd stay dry. He dozed off until water, rushing through the culvert, soaked him.

The next day he developed a fever. When he came to a small town, he went into a drugstore and asked the pharmacist if he could get something to bring his temperature down. The pharmacist advised him to see a doctor.

"I can't," Joe said. "I have no money."

The pharmacist asked if he was a veteran.

"I guess so. I was in the navy till the war ended."

"Then you're eligible. There's a veterans hospital across the river in Kentucky." The pharmacist paid for Joe's cab fare.

Joe was so sick he couldn't think straight when he entered the hospital. A corpsman took his temperature, which was 106 degrees. The corpsman must have been new, because he directed Joe to a chair until the doctor was ready to see him. He didn't know how long he sat there until a doctor picked up his chart, saw the temperature, grabbed a thermometer, shook it down, and stuck it under Joe's tongue. While the temperature was recording, the doctor was moving nervously from one foot to the other. When he saw the 106 recording, he berated the corpsman. "You idiot. This man has pneumonia. Get him in bed fast!"

Within minutes, Joe was surrounded by nurses and doctors doing all kinds of things, including soaking him in cold wet towels to bring the fever down. The only antibiotic available then was penicillin, so they gave him that. If he had not got to the hospital when he did, he probably would have died by the side of that lonely road.

While recovering, he wrote his father and asked for enough money to take a bus home. Somehow his dad got the money. Once he was back in Carlsbad, Joe finished his apprenticeship.

He was 23 years old when a friend introduced him to a girl who had recently moved to Carlsbad. Mary was the oldest of a large and very poor family from Oklahoma. Joe began dating her and they were married a few months later. Two years later, their first child was born. They named her Janet. In another two years, Tommy was born, followed in two more years by Terry. This third baby had breathing problems that required more care. During this period, their marriage hit many rough spots, and work in Carlsbad was slow.

When Joe heard of a job in Utah building a uranium plant with overtime pay, he went by himself to check it out. When he determined the job would last, he bought an old 27-foot Kozy Koach trailer and moved his family to the jobsite. In 1957, almost everyone on the job lived in tow-behind trailers.

Joe and four other electricians decided to build a raft and haul it to Bluff, a town 25 miles away by road and 40 miles by the San Juan River. They pictured themselves floating down the river for 40 miles and coming out at the Mexican Hat jobsite. It sounded like fun.

River rafting was not common at that time, and none of the five knew anything about it. Their plan meant building a homemade poor-man's raft with material scavenged from the area. Their biggest mistake was building the raft from heavy wooden timbers with an empty 50-gallon oil drum wired to each corner.

"We rescued 10-foot conduit poles to keep ourselves in the deeper part of the river," Joe explained. "We all had an inflated inner tube around us in case someone fell in the water. We brought our own lunches in paper sacks, and I had my stainless steel coffee thermos. We also had a cooler filled with beer."

Very early Saturday morning, they loaded the raft on a pickup truck. That was when they realized how heavy it was. With the adventure already underway and complete confidence in their ability, they drove to Bluff to launch the raft. It was a struggle getting it off the truck and into the water. They managed, with help from friends who had driven over to see them off and drive their pickup back.

They tied down the cooler of beer, in fear it might fall off, and piled their lunches next to it. Joe kept his thermos that still had coffee. One person sat on each oil drum and the fifth man sat in the center of the raft. They waved good-bye to their friends, who were promised they could use the raft the next weekend. As they drifted into the action part of the river, Joe thought, "What a wonderful adventure this will be."

During the first couple of miles, they told each other what fun it was. Someone suggested they take the raft again next weekend and make their friends wait. Everyone agreed.

Suddenly the current began pulling them straight to a rock cliff. The joking stopped. They fought to steer the raft away from the cliff while the current drew them closer. The current won. When the front of the raft smacked into the cliff, the current pulled the back under water, throwing them off.

When the raft came up, it was upside down, and five men were bobbing down the river, pulled by the current well ahead of the raft.

They managed to swim to one side of the river and climb out, staring at the turbulence. The raft was nowhere in sight. It was their transportation for the next 40 miles.

Then someone spotted the raft, and one man swam out to get it. Maybe they unconsciously anticipated something happening, because there was a long rope coiled and tied on the raft. The man who reached the raft threw the rope to one of the guys, who waded out as far as he could. He grabbed the rope, and with everybody helping they were able to beach the raft.

It was upside down. Their lunches and Joe's thermos were gone. It was impossible to turn the raft right-side up, so they climbed on and continued down the river. It soon became obvious they could not control where the current took them nor could they stay in the deep part of the river. As the raft traveled on at a swift speed, they worried what would happen if they hit one of the large boulders that lined both sides of the river.

They soon got their answer. When the raft hit a boulder, the current flipped it over again. All five were thrown off and had to swim to get back to the raft. Now that it was right-side up, they could get a much-needed beer out of the cooler, which they had tied down so well that it had survived two wrecks.

Joe said, "This wasn't such a great idea. Someone could get hurt bad."

They decided the next time it looked like they might hit a boulder, they would jump off and swim back to it. After a few more miles, someone saw something shiny bobbing up ahead. "Hey, Joe, isn't that your thermos?"

Joe jumped off and retrieved his empty, but still precious, thermos. He put it inside the partially empty cooler.

They hit several boulders throughout the day and usually the raft turned over. They found it was easier to jump off and catch the raft. When they were in the water, they traveled at the same speed as the raft, so it was easy to grab hold and climb on whether it was upside down or right-side up.

They worried about reaching Mexican Hat before dark. The safety procedure they were using was difficult in daylight but would become dangerous after dark. It was getting dark when, with a sigh of relief, they completed their trip.

They had long since decided one trip was enough; they would honor their promise of lending the raft to their friends. Joe was worried. Other electricians were organizing raft trips for every weekend. Sooner or later someone would be hurt or maybe killed. So when they reached the pier at Mexican Hat, Joe said, "We should get rid of this raft. If people continue to use it, someone will get hurt."

"We can't do that. What will we tell the guys we promised to let use it? They helped us find parts and launch it."

Joe said, "We'll tell them the truth. We may have saved their lives by letting the raft go."

They watched it bobble on down the river with no one on it. Joe said, "It's a miracle we all made it back without anything but bruises." Once again, he had looked death in the face and beat the odds.

The summer passed swiftly. After spending two winters in Alaska, Joe would not subject his family to a winter in Utah. Electricians don't *quit* a job; they "drag up." *(Years later, when Joe went back to Mexican Hat, he learned that everyone who stayed there after the plant opened had died of some kind of cancer. The uranium plant was torn down and covered with a huge pile of dirt and concrete. That story never made national news.)*

While towing their 27-foot trailer to Southern California, Joe realized it was not designed for travel. The tongue weight made it difficult to maneuver, and the trailer was too heavy. It was built to stay in one place. They sold it for the down payment on a house and settled in California to raise their children.

Their marriage had financial problems in spite of his good wages. "Mary insisted on handling the money and paying the bills. The problem was, she didn't pay the bills." Joe said. "I don't know what she spent the money on. I only know we had two cars repossessed and lost the house because we were months behind in payments."

Mary wanted to go back to work as a waitress. They were so far behind on bills that Joe agreed, hoping they could catch up.

Before long they had a fourth child, another boy, named Timothy. It soon became evident Mary was more interested in what happened on her job and in the lives of people she worked with than with him, her home, or her children. She worked evening shift, so Joe fixed supper and put the children to bed.

Even with Mary making money and Joe working steady, they were always in debt. They were fighting more often. Still Joe convinced himself it would work out. Mary was his wife, the mother of his children, and he was in love with her.

Mary didn't feel the same. She announced one day that she was in love with a police sergeant she met at the restaurant and was going to marry him. When Joe learned who it was, he said, "He's already married and has kids."

She replied, "We both agreed to leave our families and start a new life together."

Joe said he should have seen it coming but refused to consider such a thing. His consolation was that she left the kids with him. She didn't attend the court hearing and signed over custody of the children to Joe. The judge couldn't understand it, but Joe realized she wanted to be free of responsibilities.

Back: Janet and Tommy; Front: Terry (left) and Tim

Carla, Mary's younger sister, lived nearby with her husband and two little boys. They moved into Joe's house until Joe could hire a live-in housekeeper. With the aid of a series of live-in housekeepers, Joe and the children did fine, and he paid off all the debts Mary had left them.

The kids never asked where their mother was. If she missed them, she never called to ask about them until three months after they had split up. When she did call, she asked if she could visit the kids.

Joe said, "Of course. When do you want to see them?"

"In about 30 minutes," she replied.

When she arrived, Joe went into his bedroom to read and let her have the front room to visit with the children. She had only been there 15 minutes when she came to his door. "Can I talk to you?"

He put his book down. "What do you want to talk about?"

She admitted the policeman she was in love with never intended to leave his family. "Don't you think I deserve another *chance*?"

"You sure do, but not with me," Joe said. He was making a determined effort to put her out of his mind and life.

Yet, he was lonely. He wanted to find someone to help raise his children and give him the companionship he craved. When he heard about an organization for single men and women, called Parents Without Partners, he decided to check it out.

First he went to their monthly dance. He was standing alone, feeling awkward, when a little petite lady introduced herself as Kay and welcomed him with a smile. If all the people were as friendly as she was, this PWP organization was exactly what he needed.

They went for coffee after the dance and he was impressed with her until she told him she was quitting her job and taking her savings to go on a camping trip for the summer. With so many women to pick from, he wasn't going to get involved with another irresponsible person.

One reason there were more women than men in the group is that men don't usually have custody of their children and are not inclined to join clubs. Many weren't interested in meetings or discussion groups but came to the monthly dances.

From the beginning, Joe was popular because he was fun, and at the discussion groups he listened as embittered women released their frustration about former husbands. Most men don't listen.

Joe started a committee called Fathers-at-Large. Once a month, he took all the boys who had no contact with their fathers on some kind of outing. As the group grew, Joe got a friend to help him.

Someone requested that girls who didn't have contact with their fathers should be included. Joe agreed and got a third man to help. The kids loved the outings, and it also helped the mothers. Joe soon became a member of the chapter's board of directors.

Joe had been single for over two years and had dated a number of women, some of whom obviously wanted to marry him. He *wanted* to be married, but somehow no one seemed to quite fit what he was looking for, and he didn't know why.

CHAPTER 9

Second Chances (1964)

A NICER HOUSE became available for rent two blocks from ours. Cathie pleaded with me to rent it. I could afford it, so we moved in. The boys had one bedroom and Cathie and I shared the other one. We were moving up in the world. The kids picked the colors and helped me paint the inside. I was able to buy used furniture. It was the first house I lived in where all the furniture belonged to me. Now I had monthly bills to meet, but with one full-time job and one part-time, I made payments ahead. I was amazed at how much further my money went with no bar bills to pay.

Joe invited me to go with his group of friends to a beach party to see the "grunion running." I don't know if he was between romances or if he recognized my interest when it was discussed at a meeting. Whatever his reason, I gratefully accepted. I liked him, but I also was eager to learn more about this annual event that happened three blocks from where I had lived for over two years. You miss a lot when you work all the time.

The grunion is a tiny fish with a silver tail it uses to dig a hole in the sandy beach. The female digs the hole; then the male raps his body around her to fertilize the eggs as she lays them in the hole. Mating only happens once a year on a predictable day.

Joe took me home, kissed me goodnight, and then left to join the rest of the party group who went from the beach to someone's home. I had to hurry to get to work.

When Joe started Fathers-At-Large, my boys looked forward to each one. They were especially excited about a trip to an airport. When Joe opened it to girls, Cathie was elated. Both Joe and his main helper became her heroes. She wanted to have a father like them.

In the meantime, I was appointed camping chairman. Once a month, we went to a nearby state camping area. My kids loved it. At first we had good turnouts, but soon the other women and their kids, for whatever reason, stopped coming. Then on one trip I was the *only woman*, when three guys showed up: Joe with his kids, and two others who didn't have kids with them. My kids and Joe's had a great time with three father figures. Joe taught me to play chess, and I actually beat the other two guys—but not Joe. After I announced the outcome at the next PWP meeting, our camping trips were better attended.

In early 1964, Frosty, the chapter president, asked me to run a weekly discussion group for newcomers. Older members complained that newcomers monopolized discussions to disparage ex-mates. The board (of which Joe was a member) decided new members needed to attend six newcomer discussions before they could attend regular ones. My job was to help them express their anger and hate so they could move past their bitterness.

The first meeting went fine, but at the second meeting a newly divorced police officer dominated the meeting, verbally attacking me as well as other women. He reminded me of Jim, and my old fears came back. Nobody got anything of value from that meeting.

The next day I called Frosty to explain what happened. "I'm not the right person for this job. Angry, dominating men scare me. They bring back too many bad memories."

"You *are* the best person to do this. You're empathetic and a good listener. If I find a backup to control people like the police officer, will you try again?"

Reluctantly I agreed, not knowing the person he would ask to help me was Joe Peterson.

On the way to the meeting, I said, "I probably sound silly, but my husband was an abusive alcoholic. He only beat me once, but it was really bad, and he constantly threatened to do it again. This policeman reminds me so much of him that he scared me. I just can't handle aggressive men. That's why I wanted to quit."

He reached over and took my hand. "You don't have to worry. I promise you. He won't give you any more trouble."

When we arrived, the policeman was there. A few seconds later, both Joe and the policeman were gone and did not return until I called the meeting to order. I never knew what Joe said, but that night the policeman was very quiet. He did not harass me or anyone. I never saw him again.

After the fourth meeting, on the way home, I said, "I guess I don't need protection anymore. I don't think that guy is coming back."

"I didn't think he would. He realized this group wasn't for him. I enjoy the discussions and like the way you talk to people who are upset. I want to continue. You never know when another aggressive woman-hater will show up."

We developed a routine. I drove to his house, left my car there, and we drove the rest of the way in his car. His house was halfway between mine and the meeting place. These were not dates, but they gave us an opportunity to clear up misconceptions we had about each other. He was dating someone else on a steady basis.

Frosty decided to start a special group of six men and six women he thought were ready for more in-depth discussions. It was by invitation only. Joe's current girlfriend was not part of the group. Because the group was the same each week, we were more open and honest. Still, there were things in my past I wasn't ready to share.

These special discussions included religious differences between partners, expectations of a new marriage, our children's reactions to our dating, and problems with stepchildren.

In early September, Joe had hernia surgery. Whenever members or their children had surgery, it was my practice to get that day off. When they came out of surgery, I would be there to "special" them for the next eight hours. When Joe's surgery day came, I did the same thing. Since I wasn't sure if I could get that day off, I did not tell him ahead.

When he returned from the recovery room and saw me in my white uniform and cap sitting by his bed, he looked pleased. Since he was in a four-bed ward, I helped the other guys when Joe was sleeping. During the night, a nurse came in to take his blood pressure and pulse. He took her hand and kissed it.

"What was that for?" she asked.

Joe laughed. "Sorry. I thought you were someone else."

The next morning, the guys teased Joe about his "cute private nurse." Every day after work, I visited him while he was in the hospital. I'm sure he had other visitors at regular visiting hours, but my uniform gave me special privileges.

We had been to so many discussion groups together that we never ran out of things to talk about. My hair was getting really grey—so I had been coloring it brunette. The shades changed depending on the hair dresser, so it was evident. It no longer bothered him. I guess it was just that the blonde was so fake-looking.

Joe and his last dating partner had agreed to start seeing other people before he went for surgery, so maybe it was the timing. After he went home, almost every evening I visited him. We played chess, with him propped up by pillows and me sitting cross-legged on the foot of his bed. I was determined to win *one* game.

We already knew each other's children from camping trips and his Fathers-at-Large excursions. Our kids got along well. Joe and I had been friends for over two years, so it was easy to talk openly about many subjects, including more personal things than we talked about in our special 12-person group. In that group some people had begun dating each other. I didn't think of my evening visits as dating; there was no kissing or even hand-holding involved.

Joe's doctor told him he could go back to work. That night when I arrived, he was sitting in the living room with his kids, playing a card game. I joined in. When the housekeeper called the kids for bed, we continued sitting there instead of playing chess.

This time our talk was more about our previous marriages. I felt I could confide in Joe the way I had Peggy years ago, with the instinctive knowledge he would keep it to himself. I told him more about things Jim had done and my fear of his taking the kids from me until I could get full custody. I told him what the Albuquerque police had said about Jim's rights. Until my divorce was final, I would not rest easy. I think that was when we started holding hands.

He responded by telling me about his childhood struggles with his family and his education. These were things he had not talked about in discussion groups.

I'm not sure how it started, but we made love. It was late when I said I should go home. Always before I just left, but this time he walked me to my car. I got in the car and then he leaned down and kissed me.

"Move over so I can sit down," he said. We kissed a few more times, more passionately, and then he blurted out, "Will you marry me?"

It was a shock. We hadn't even really dated. "There is one more thing you need to know before you ask that."

Then I told him about Sherry, my child that no one at PWP knew about. I told him I had her before I was married, that she had run away, I had no idea where she was, and I felt I had failed her.

"Is that it? If it is, I'll ask you again. Will you marry me?"

I knew right then I had found my soul mate.

The next morning I told my children I was getting married. When they heard it was Joe, the boys were pleased, but Cathie was jumping with excitement.

"*I* did it," she announced. "I've been praying every night for God to find someone to take care of you and be my father."

That was October the sixth. It would be an income tax benefit to marry before the end of the year. I said, "My divorce is final December 18. Wouldn't it be fun to get married that same day?" When we found it was a Friday, we could have a weekend honeymoon without missing any work.

We both had a lot to do. He had tickets to a musical with his last dating partner. He wanted to tell her before she heard from some other source and also to give her the tickets so she could still go to the musical with someone. It had been his idea to end their relationship and be friends; she was still hoping they would get married. He wanted to make it as painless as possible for her.

He also had to resign from the PWP Board because, once married we could no longer be members. At the next board meeting, he made his announcement, and they were shocked when he said he was marrying me.

Kay and Joe at a PWP outing in 1964.

"I didn't even know you two were dating!" someone said.

"We started seriously dating in September," he said.

Frosty said, "I always thought you two were a perfect match." Had he been manipulating us from his president's position? I always did like Frosty. I did not realize Joe considered my nightly visits as dating, but it made our sudden engagement sound more logical.

Since we were combining two families, and both of us were currently renting, we agreed it would be easier on the kids to buy a bigger house and start on common ground. We found a new four-bedroom house in a small housing tract in a different part of Huntington Beach.

Although he had paid off all his debts, Mary had ruined Joe's credit; the loan officer required he have a co-signer. We weren't married yet, and I had excellent credit. I co-signed as his cousin. The loan officer knew it wasn't true, but his concern was getting paperwork his boss would approve and his commission depended on.

We purchased the house in November, so Carla and her family moved in to take care of Joe's children until I moved in during Christmas school vacation. One day, to be funny, I asked, "Are you marrying me because you are tired of having housekeepers quit?"

He retorted, "I didn't think you'd catch on this quick." One of the things everyone liked about Joe was his ability to laugh at himself and life.

I went to the courthouse early on December 18 to pick up my final divorce papers. Joe met me there on his lunch break to get our marriage license. The ceremony was set for after work that same day. I was legally single for about six hours.

It was a short ceremony in the minister's upstairs office. Joe's best man was Frosty. My bridesmaid was Carla. Her husband was the only other adult along with seven grinning children standing beside us.

Another PWP friend loaned us the use of her desert cabin for our honeymoon. It was a perfect place and my introduction to how beautiful the desert can be. The plants were different from many of his boyhood, but Joe knew their names and why each was unique.

I wished we could have stayed longer, but we had a family to raise and jobs that only paid you when you worked.

CHAPTER **10**

Uniting a Family (1965—1970)

THE NEW HOUSE was a long way from the high school where Skip's friends went, so rather than change, he rode his bicycle to his old school. The other kids were still in elementary school, which was two blocks away. The seventh and eighth grades were combined in one classroom, so Janet and Scott both had the same teacher. Cathie and Tommy were in the same grade but had different teachers. Even Tim, just turning five, would be able to walk to the school because it had a kindergarten.

There are adjustments after any marriage, but when two sets of children brought up differently are added to the equation, it becomes a challenge. Since Joe had live-in housekeepers for two-plus years, his kids didn't have regular chores; even making their beds was the housekeeper's job. My kids always had chores, so new rules were made that applied to the whole family. Allowance money was distributed only if chores were completed.

The children would all make their own beds, keep their room picked up, and take dirty clothes to the laundry. After meals, they carried their plate and utensils to the kitchen. The one assigned to supper dishes rinsed the dishes and placed them in the dishwasher for final cleaning and sterilizing. This job included washing cooking pots and pans. Someone else was assigned to emptying the dishwasher and putting the dishes away.

These jobs changed every week for the five older kids. Terry, with Tim's help, was in charge of the trash. Some kids did their chores without complaint, some complained as they did it, and some paid someone else to do it for them. We let them work it out themselves.

Statistics for a marriage surviving the raising of a combination family were not good. Of our PWP friends, all who tried it ended in divorce or gave up custody of one set of children.

Every newly married couple has a period when they are adjusting to a partner and establishing a bond. When you add stepchildren to the mix, they resent having to share their parent with a stepparent. When a stepparent tries to discipline a stepchild, the siblings will usually rise to the defense. Also, it is a parental instinct to defend your child when arguments arise. We were determined we would not let any child come between us. Our children would be with us for another four to 14 years, while we intended to be together for the rest of our lives.

Joe and I made a pact. When possible, the parent would dole out punishments. Kids learn quickly. If one parent doesn't give the answer they want, the child will go to the other parent hoping for a different answer, and it can end up with parents arguing. Our pact also included that, if either of us had given an answer, the other parent would back it up regardless of personal feelings.

If we felt the decision should have been different, we discussed it later when we were alone. When the kids discovered we would not contradict each other, they usually figured out which parent was more likely to give permission for a particular request.

We learned a lot through discussion groups and especially the 12-person group where we discussed problems that could arise with stepchildren. Now Joe and I helped start a group called Remarried, Inc., a continuation of PWP for people who now had partners. Our primary emphasis was discussion groups in which we shared new problems. It became a national organization several years later.

We also attended community college lectures on parenting that were helpful and read books. One was on using "natural consequences" for punishment. For instance, if a child forgets to bring lunch to school, most mothers rush to school so the poor kid wouldn't go hungry and then lectures the child later. If our children forgot, they went hungry or had to beg part of a friend's lunch. They seldom forgot more than once. This method was recommended by a psychologist. We used it whenever possible.

We held weekly round-table discussions that everyone was required to attend. This was when they could bring up their grievances regardless of the subject. It was also a time for the family to agree on chores and pick a date for our next camping trip.

At one of these meetings, Cathie brought up a problem she had with Janet. The two girls shared a room and a large bed, which was in the center of the room for access from either side. Cathie always hung the clothes or took them to the laundry and made the bed. Janet didn't get around to such trivial things.

Cathie said, "I'm tired of being the one who always changes the sheets, makes the bed, picks up our clothes and keeps our room clean. Janet doesn't do anything. I don't think it's fair."

We traded their big bed for two twin beds and put one on each side of the room. They had to work out the clothes and cleaning themselves. Cathie's solution was to draw an imaginary line down the center of the room. Her side was always neat as a pin. Instead of worrying about Janet's clothes, she tossed them over on Janet's side of the line. I can't say they ever enjoyed sharing a room, but they stopped fighting about it.

Joe and I also had our own discussions. The first major one was about who would handle bill paying. Both of us had come out of marriages where our partner used our paycheck to pay the bills, and neither handled it properly. Now that we had handled our own bill paying, neither of us was willing to give the control away.

We compromised. Every Friday evening was set aside for paying bills. We scheduled the bills so that a paycheck was devoted to each one. My plan was for us to sit down together and make out the checks so we were confident they were paid. It worked for the first month or two. By then Joe, who felt it was a chore, trusted me to do it while he watched TV with the younger kids. Everything was paid ahead, and we had a savings account started.

Another agreement we made was Joe's suggestion. We would never, under any circumstances, lie to each other. If one of us lied and the other found out, it would be impossible to fully trust that person again. I readily agreed and neither of us ever did. I remember a time when Joe dropped me off at the hairdresser while he did errands. He got distracted and forgot to pick me up. When he arrived, I assumed it had taken longer than we anticipated, but to be funny I said, "What did you do, go home without me?"

It would have been easy to lie, but he didn't. Rather sheepishly he said, "Yeah. I forgot I was supposed to pick you up." We both laughed.

We argued but never "fought" about anything—neither of us ever called the other by an ugly name, brought up past issues, or even raised our voices. How did we settle an argument when we didn't agree? We made another pact.

If we wanted to do different things or even listen to a different TV program, we'd each write down on paper a number from 1 to 10 that determined how important it was to us. When we showed our numbers, which had to be our true feeling, the higher number won. We did this throughout our marriage, although we soon abandoned writing the numbers on paper. It avoided arguments because each knew the other was telling the truth. If it meant more to the other person than to us, neither of us minded giving in. Almost never did either of us give it an importance of 10.

With seven children all wanting individual attention, we had little time to enjoy each other. To solve that, we set Saturday morning as our time. We left the kids to fix their own breakfast while we went to breakfast either by ourselves or with the couple who were our best friends.

When a child was having a birthday, we took only that child to Saturday breakfast with us. This enabled them to have special time with us. The first year or two, everyone took advantage, but as my boys got older, they sold their breakfast to Cathie, who loved these times more than the others did.

The first few years, camping was an important part of bonding as a family. All too soon, my two boys were tired of camping and wanted to stay home and do things with friends. This made room for the younger ones to bring a buddy along, so we always had a gaggle of kids.

On his way home from work, Joe spotted a tiny teardrop trailer. He called the owner, and we met him to see the inside. It was ideal for us. We were using a tent and sleeping bags on the ground. Joe's back always bothered him after these trips. Even a fold-up cot didn't help.

This little trailer had nothing but a bed—with a real mattress—inside it. The back end lifted up into a cover to shield a place for a kerosene cook stove and a little sink to wash dishes in, plus a small storage space for groceries. The sink emptied into a bucket we placed underneath it.

When we were traveling, we piled the sleeping bags, tent, Coleman lanterns, fishing poles, and other equipment on the bed. This gave us more seating in the car. When all seven kids were with us, we took out the back seat, making more seating space including the trunk.

A few years later, we traded the tent and teardrop trailer for an Alaskan camper mounted on the back of our pickup truck. This gave us more traveling space, more storage, and added comfort.

In the summer of 1965, we had been married for about six months when Joe's mother and new stepfather, Walter, drove from New Mexico to introduce Lena to his family and for Lena to see Joe and her grandchildren. I was in the kitchen when Joe answered the phone and whispered to me that it was Lena calling from Los Angeles. From the conversation, I realized she was asking him to come and get her.

I was shocked to hear him say he wouldn't drive to Los Angeles. He gave Walter's son directions for getting to our house.

"Why wouldn't you get her?" I asked when he hung up.

"I don't care if she comes or not. If she got this far, she can get the rest of the way. You'll understand when you meet her."

I remembered the embarrassing things she did when he was a child, but this wasn't the Joe I knew who went out of his way to help *anyone* who asked.

I was eager to meet my children's only grandmother now that Grammy was out of our life. I pictured her as being small, white-haired, and pleasantly overweight. It took my breath away when a tall, broad-shouldered lady with flaming red hair grabbed me in a bear hug. She was wearing shorts that stopped at her crotch. At first I thought it was her underwear.

Walter had one eye with a patch over it and the other covered with glasses that looked like the bottom of a coke bottle. Evidently Lena did all the driving. Walter's son said sarcastically, "I'll come get her in a week unless you can find time to bring her to my house."

"Oh, I'll find time to bring her back." Joe laughed.

Lena began talking about how nice the police were in Arizona. (Red lights were made for other people, not Lena.) "We were crossing a street in Tucson, when a big truck blared his horn at us. He musta been goin' fast 'cause his brakes were screeching. Nearly ran us over. He was mean, too. Stopped in the middle of the street and shook his fist, so I gave him the finger."

"You think maybe you did something wrong, Lena?" Joe asked.

She ignored him. "Wasn't long after when another policeman stopped us just to ask how we were doing. I told him we'd just got married and were going to California to meet each other's families. He told me to drive careful. Before we got to Phoenix *another* policeman met us. He escorted us all the way through Phoenix with his lights flashing. Walter thinks a second police car was following us."

We decided to take Lena to dinner. Joe was driving, I sat in the middle, and Lena was by the door in the front seat. Five kids were crammed in back. As Joe started the car, Lena reached over and pinched my breast. Joe said, "Lena! What are you doing?"

"It's okay, honey," she said. "I wanted to see if they were real."

I told myself to be prepared for an interesting week.

That was my introduction to my children's new grandmother. My boys were puzzled but ended up enjoying her antics. It was obvious Joe's kids adored her, and Cathie was fascinated. In spite of her age, Lena did everything with the kids—bicycling, skate-boarding, running races and, of course, buying them candy when they walked down to the corner liquor store for her to get "something to drink." Water, soda pop, milk, and juice were all we had in our refrigerator. Joe normally had a drink when he got home from work but didn't drink that week and had hidden the liquor.

No matter what I cooked for meals, her comment was, "You can't even cook, can you? I'm no cook myself, but I do better than this. Mary was a good cook."

She never called me Kay. I was "you" or "her" the few times she referred to me. I asked Joe if he knew why she didn't like me.

"She doesn't like any women. She talks bad about everyone including her own daughter."

I was relieved when her visit ended.

The summer between his sophomore and junior year, Skip couldn't find any job except delivering give-away newspapers on the weekend, partly because he was small for his age. He was interested in what Joe spent so much time on—repairing our cars.

"Do you know how a car works?" Joe asked. When Skip shook his head, Joe said, "Climb up here beside me, and I'll explain it to you."

Joe was amazed at how quickly Skip caught on and how interested he was. "I'll show you how to make money this summer, but it'll be hard work. I'll *loan* you the capital to get started."

They drove around the neighborhood searching for a car that had not been moved for a while. Tires would be flat or it was evident the street sweeper was going around it. The first time they found one, when they went to the house to ask if it was for sale, they were told it was broken down. Joe paid $25 for it.

Skip had all summer to work on it. Each night Joe would instruct him on the next step. Skip listened and absorbed information as if he were a sponge. Usually that day's assignment was finished and Skip was impatiently waiting for Joe to tell him what to do next. At the end of summer, he sold it for several hundred dollars and had learned a lot.

After he paid Joe back the loan, he had more than enough to do another car. He was on his third car when he was old enough to get his driver license. This time he bought a nicer broken-down car. Joe said, "When he finishes this one, he'll want to keep it."

He was right. Now delivering papers was easier. He hired Terry and Tommy to fold the papers and later hired Terry to ride with him and throw the papers from the car.

Scott became involved in tennis and played every afternoon. This meant he had to walk three miles. When he got home, he laid out a whole loaf of bread in a double row. Then he put peanut butter on one row of bread and jam on the other. This "snack" plus a big glass of milk held him until supper.

Everyone was expected to be at the table at suppertime unless they had permission to go somewhere. Supper was a time for the kids to tell us about their day and their friends. The rule was you tell Joe how much food to put on your plate, and then you eat it *all* or no dessert. One night when we were having spaghetti, Scott kept holding his plate out silently asking for more. Joe had already piled it very high.

"OK, smart boy," Joe said. "But you better eat every bit of it." Scott grinned. Before long, he was holding out his plate again. "Can I have more, please?"

That same night, Tommy fell asleep sitting up at the table. Joe tapped him on the back of the head and Tommy's face fell forward into his plate of spaghetti. He sat up with spaghetti sauce dripping off his chin and wondering what everyone was laughing about.

At Christmas, construction work slowed down. This happens periodically, so Joe prepared for it with emergency savings. During these times I would work vacation relief for various nursing homes and hospitals. The previous Christmas, we had gone all out on presents. At our round-table discussion, I said we wouldn't have as many presents this year because of Joe being out of work, so to make a list of the three things they wanted most and they would probably get one of them.

At that time offices used IBM cards with holes punched in them for their records and then threw them away. A friend showed me how to staple them together to make a wreath. After it was spray painted and a few ribbons plus a cheap ornament added, they were attractive. I told the kids they could help make these and take them around the neighborhood to sell. With the money they made, they could each buy a present for the person whose name they drew.

Tim took his wreath to the Schmidt house where his best friend, Jeff, lived.

When Jeff's father saw the wreath, he asked why Tim was selling it. Tim started crying, "Daddy's outta work and we won't have *any* presents unless we sell these wreaths."

Andy bought the wreath and said he wanted more. Andy was a cook at a big restaurant, so he decided to help. He put the wreath on a display stand with a sign:

Family of 7 children
Father out of work
Buy a Christmas wreath to help them

People started ordering wreaths. When Tim brought the news home that Andy was selling them at his restaurant, we started making wreaths like crazy. A small family project became a thriving business. For a short time.

One day while we were making wreaths, a policeman came to the door. One of the kids let him in and he came to the table where we were working. He asked Joe to step into the other room with him.

He wanted information on the kids so the police fund could buy presents. Joe explained we didn't need help and to use their fund for a family that did. Unconvinced, the policeman tried to get Joe to let us be on his list. Joe offered to show him our bank statement to prove we had money. He finally crossed us off his list.

A few days later, we got a telephone call from a charity that wanted to provide our Christmas dinner. Again Joe explained we had the money we needed until work picked up. The lady told him his pride was getting in the way and it wasn't fair to the children. Joe asked who told her we needed help.

"One of your neighbors reported you have seven children and are out of work. That's all I can tell you."

Joe confronted the kids. Tim admitted he told Andy, if they didn't sell wreaths they wouldn't get presents. Joe called Andy, thanked him for wanting to help and explained our situation. Andy took his sign down, bringing an abrupt end to our wreath-making business. We had enough wreaths already made to hang on every door inside and outside the house.

The summer of 1968, Joe and I took a month off and traveled from the west to the East Coast and back by a northern route in the Alaskan camper. Joe was opposed to the idea until I showed him I had been putting the money I earned into a savings account. I showed him that our bills were paid a month in advance. We had more than enough for our trip expenses, plus enough to pay Carla to move in with her family and take care of the kids. He was surprised and agreed to our second honeymoon.

It was my first chance to experience the wanderlust that for years I had pushed below the surface. The thing I remember most about that trip was the thrill of discovering what lay around the next bend.

We had one incident on the way home that we often laughed about. We were in San Francisco at the peak of the hippie movement and tail end of the Haight-Ashbury days of free drugs and anything-goes behavior. We parked and watched the action around us. The colors and strange clothing people wore were intriguing.

On the surface, they were having fun, but it was not genuine; it was the hallucinating effect of drugs. Both marijuana and a drug called LSD were popular at that time. We watched one baby being passed back and forth between men and women who used the wide-eyed child as a begging chip to get money. That disgusted me. I suggested we look for a place to park overnight.

The map showed a hill with a scenic viewpoint on top. When we drove up to see if there was room to park, we found it deserted. The view of San Francisco was amazing. What a great place to spend the night even if it became a lover's lane later. That had happened before, but lovers never bothered us. Joe always made me feel completely safe.

After supper we were playing chess when the police drove up. Two policemen approached us. "Good evening. Are you planning to spend the night?"

Joe explained we'd been on vacation and were headed back to our home south of Los Angeles. "There's no sign about parking so we thought we could spend the night here."

"Do you know you're in the middle of the Haight-Ashbury District?" one policeman asked. "In another couple of hours this hill will swarm with drugged hippies. We had 164 unsolved murders here last year and aren't trying to beat that record."

Wow! I started putting things away while Joe got directions to a nearby, well-lighted place on the beach where we could stay. When we got there, we found several campers and a trailer spending the night.

In September 1968, Scott started his senior year and had signed up for a full load of honors classes. By the end of the first week, he felt overwhelmed.

On Friday night, he and Harry, one of his three closest friends, were drowning their sorrows in beer. Another of their friends had moved to Oregon. Harry started crying. He was engaged but not ready to get married and wanted to get out of town but didn't want to go alone. Scott made the brash decision to go with him and escape the overload of classes. They went to John's house, woke him, and told him they were on their way to Oregon to see Kris. John decided to go with them.

They didn't bother to tell their parents, so I spent some sleepless nights until Scott called and said they were "off on an adventure."

In Oregon, Scott and John worked for six weeks in a cannery. Harry had already come back to his fiancée. John wanted to go to Chicago to see a girlfriend who moved there, so Scott went with him. John's dad found out they were there and had John arrested and brought home. This left Scott alone.

Scott stayed in Chicago only long enough to build up a cache of money by working long hours at a factory. Then he flew to Miami in search of Grammy. Scott said, "I remember the romantic feelings of adventure and danger as I flew to Miami, then got a bus to the suburbs where I had lived nine years earlier."

He walked for miles until he recognized the names of streets. He found her in the phone book. He stayed with her a couple of weeks and then talked to his father on the phone. Jim was in Massachusetts and begged Scott to come there. Grammy's parting words to Scott were to keep his money a secret from his father.

Jim lived with a man and his mildly retarded teen-age son. The son's sole friend was a boy a few years younger than Scott but very bright. Scott said, "I remembered his parents, Carol and Jerry, from childhood. Dad said he was actually my half-brother. He also said he had more children and another failed marriage after ours."

Scott stayed with his father for a few weeks. They played cribbage or penny ante poker while drinking whiskey and water. "I developed a taste for it—not to mention endless cigarette smoking," Scott said. "We often went to bars. Dad bought tiny bottles of vodka at the liquor store. I was 17, so I ordered orange juice and Dad sneaked the vodka into it. I got bored going to bars, but when he went by himself, he came home so snookered I hardly recognized him.

"I'm glad I saw him again before he died. He was an intelligent man who didn't belong and didn't want to belong. I didn't feel sorry for him. His life seemed to suit him."

In January 1969, Scott returned home, and went back to finish his senior year. He started using drugs and was drinking a lot until a psychologist convinced him that his genes were such that he would become an alcoholic and probably become addicted to drugs. The memory of his father helped him realize that wasn't the life he wanted. He stopped drinking and using drugs, married, became a heavy equipment operator and has worked for the same company for 25 years. He had a daughter and a son, both of whom went to college.

Sherry got in touch with me through her grandfather when she turned 18. By then Joe and I were married. Her first husband turned out to be an abusive drinker. We sent her money to move to California with her toddler and helped her start a new life. She changed her name, now has four living children, eight grandchildren—one of whom she and her husband are raising—and one great-granddaughter, which made me a great-GREAT grandmother at age 80.

The summer of 1969, Janet wanted to stay with her mother for the summer. She ended up living in New Mexico most of her life. Being a mother was her main role in life that started when she was a child taking care of three younger brothers. She was the only one of Joe's children who kept in touch with her mother and helped her. Janet and Joe had a special bond; they talked on the phone almost every week until she died at age 57 of cancer. It devastated Joe.

Cathie and Tommy started eighth grade together. Now the years of teachers passing him, even though he wasn't ready, caught up with him. The principal insisted he wasn't ready for high school and wanted him to repeat the eighth grade. Reluctantly, we took his advice. It was a mistake. It destroyed his already damaged self-image. I believe he suffered from the same dyslexia and attention deficiency syndrome that Joe had, but it still wasn't recognized. Holding him back only delayed the inevitable.

When we learned he was skipping school, the psychologist who was treating him suggested I drive him to school and watch him go inside. Later, we learned he walked in the front door and back out before my car was out of sight. He spent most of the school days that year at our house or a friend's house watching TV.

Skip and Scott shared nearby apartments with friends. Cathie, Terry, and Tim were doing well in school. Our children were growing up and becoming more independent.

At Christmas of 1969, my stepmother sent us a Christmas present that made 1970 a year of many changes.

CHAPTER 11

Call of the Open Road (1970—1972)

OUR CHRISTMAS PRESENT arrived the second week in January introducing the world of RVing. RV stands for recreational vehicle, which is not an appropriate name. It should have been called TH for travel home.

A few days later, while Joe was running a job, he called the union hall for another electrician. They sent a "tramp" (traveling electrician). That evening Joe told me Don had lived in a travel trailer for many years with his wife and now grown daughter. I was intrigued, so Joe arranged a visit to Don's family that weekend.

When we arrived at their small RV home, I marveled at the compactness. As we entered, on the right a couch faced a television. In the kitchen to our left were a double sink, two large cupboards, a refrigerator with separate freezer, and a stove and oven where Faye cooked their Thanksgiving turkey. Across from those was a booth and table that converted into a bed for their daughter. Beyond the kitchen was a bathroom with a flush toilet and shower on one side and clothes closets on the other. A small passageway between them led to their bedroom with more closets.

Yet the entire trailer was only 27 feet long and 8 feet wide. (Today fifth-wheel trailers go up to 40 feet long and have as many as four slide-outs that expand the width when parked.)

I couldn't stop talking about it. Joe said, "We can try 'tramping' when the kids are on their own. I know I'd enjoy it. I could work and we could see the country at the same time."

His words, along with Faye saying they began traveling when their daughter was a toddler, raced around my mind. They had come to California to spend the holidays with their daughter, who was going to college here. While visiting her, Don was working temporarily. If they could do it with one child, I saw no reason why we couldn't do it when we got down to our last two kids—Terry and Tim. That was only three years away.

I looked in the library for books on traveling in RVs. The only one listed in the library's catalogue was *Trailer Travel at Home and Abroad* by Wally Byam. He designed and built his own trailer in his back yard. When friends began asking him to build them one, he established the Airstream manufacturing company. His book only told about using trailers for vacations and the club he established for Airstream owners. It also told about a caravan trip they took in Europe and Egypt plus other parts of Africa. Wow! That excited me.

An Airstream must be the best kind, and owners could attend monthly rallies. However, the price was too high for us. I looked in the newspaper for used trailers.

Joe said, "Why are you looking now? Tim is only ten."

"It helps to get educated so we know what to expect when the time comes." We had been married five years, and he still didn't realize how goal-oriented I was.

After six weeks of looking, I found an ad for what sounded perfect. It was a 26-foot Airstream. I talked Joe into going to "look at it" even though I agreed we weren't ready to buy.

We both fell in love at first sight. Although built in 1959—11 years ago—it was in excellent condition. Another couple was looking at it when we arrived but decided to "think it over."

I *knew* this trailer was meant for *me*. I presented logical arguments about why we should buy while we were both working. It would be paid for when we were ready to start tramping. Then, as if I had just thought of it, I said, "If we buy it now, we can use it for camping trips until we're ready to start tramping."

Either my enthusiasm won him over, or he was so much in love he would do anything reasonable to please me. I was the helpmate and companion he wanted so badly and deep down he wanted to travel as much as I did.

We gave them money to hold it for us while we arranged a loan. When we picked it up in April, it was exactly three months since I first learned it was *possible* to travel and work. I was thinking about the coming summer vacation. We could travel while Joe worked short-term jobs and return when school started. Between our Alaskan camper, this trailer, and sleeping bags, the kids would have a chance to see some of our country.

When I presented my plan to Joe, he was skeptical about finding enough work to keep our bills paid. He was determined not go through any more repossessions. Then I showed him that my last year of working part-time had allowed me to pay both the trailer and house payments six months ahead. I also showed him our savings account. He was used to Mary spending her work money as well as his, so this delighted him.

I thought the kids would be thrilled. I was wrong. It turned out the only child who wanted to go was Tim. I couldn't understand it. To me it was a great gift we were giving them, but they didn't see it that way.

Life has a way of letting things work out. Terry was good friends with the Snows, an older couple down the street from us. They had no children of their own and had sort of adopted Terry, who was a good-looking redhead with a pleasant personality and always a hard worker. While selling their house, they had hired Terry to help them. In June they would move to a farm in Oregon and could use Terry's help there. They proposed we let Terry stay with them during the summer. Terry wanted to go, and finally Joe agreed.

Cathie presented her own plan for staying with her best friend. Jean was an only child and her wealthy parents adored Cathie. She spent many weekends with them. Cathie had a job at a roast chicken fast-food shop and wanted to earn more money. Skip offered to move into our house and supervise Cathie when she wasn't with Jean's parents.

Scott was working and shared an apartment with friends. Janet was with her mother. This left Tommy, who didn't want to go but had no alternative.

The first day on the road, we stopped at Blythe City Park so Joe could take a nap while the boys stretched their legs. Our first adventure began then. They ran back to tell me they saw a dead baby.

I went with them and found a young teen-age couple with a baby and toddler lying on a blanket under the trees. The parents were asleep, and the toddler was sitting quietly beside them, but the sun had moved, and the baby was lying directly in the sun. Her bottle had rolled out of reach. As I got close, I realized this baby was either dying or dead. I told Tim to get Joe and sent Tommy to a fire station I had seen when we entered the park.

My talking woke the parents. They seemed dazed. I picked up the baby and started doing mouth-to-mouth resuscitation until Joe got there and took over because I didn't have enough breath to continue. Firemen, who were paramedic responders in those days, arrived, followed by police. The firemen worked on the baby more for show than expecting results. The parents were distraught. They were just kids themselves and didn't realize how easily babies can become overheated and dehydrated.

We gave the police a report and told them we were leaving the state. I often wonder what happened to the parents. I hope some politically motivated prosecutor didn't put them through a homicide trial. It was bad enough to lose a baby and probably have the toddler taken away. When kids have babies, they don't know what to do or have enough money to do it with.

When we were on our way again, I said, "This is going to be an exciting trip. We aren't out of California yet."

Joe selected Colorado for his first job because his union agent said there was an "open call" for men. We took our time getting there so we could see as much as possible along the way.

Of our many weekend trips in Colorado, Garden of the Gods, Big Thompson Canyon, and the Rocky Mountain National Park were the family's favorites. Both Joe and I were intrigued with silver mining. My favorite was the town of Leadville where "the unsinkable Molly Brown" made her way into the book of legends.

The closest trailer park was in Lakewood, so we settled there while Joe worked on construction of a huge Kodak factory and headquarters. The permanent park residents were construction workers, mostly traveling without families. The rest were permanent summer weekend visitors from nearby cities.

To encourage people to visit, I made a sign using a picture of a coffeepot I cut out of a magazine and pasted on a poster board with a verse saying:

"The coffee is ready; no bother or fuss:
If you have the time, please visit with us."

I hung the sign where it was visible for people walking. I put cups, sugar, and powdered cream on the table under our awning and often sat outside knitting. There were always extra chairs to encourage people to stop and visit.

Joe never had a problem making new friends. As a child, he learned to talk to tourists; as he grew older, he honed that skill because his quest for knowledge was overwhelming. He was really the one who taught me how easy it is to start conversations whether in an RV Park or standing in a grocery check-out line.

One evening we were taking a walk in a residential neighborhood when he saw a tree he didn't recognize in someone's front yard. Dragging me with him, he knocked on the door. When a lady answered, he explained we were traveling around the country and he was interested in trees. He asked what kind of tree it was. She didn't know, but something about Joe's eagerness caused her to call her husband.

The man was as curious about our traveling as Joe was about the tree. Within minutes we were sitting on their porch talking. It was almost dark before the four of us reluctantly stopped exchanging stories.

Countless times I watched Joe use the technique of throwing out a few words that told people "Aha! Here is someone interesting." Many people are intrigued about traveling full-time and know nothing about the RV lifestyle. Some might dream of traveling but are unable to cut ties to home. Some are fascinated because it seems weird. The real satisfaction comes when you meet someone who secretly dreamed of traveling but didn't realize others were actually doing it. They are hungry for knowledge.

Life is about learning new things and meeting new people. One of the easiest ways, especially for men in a trailer park, is to raise the hood of their vehicle and look at the engine. An invisible magnet inside all open hoods draws men from every corner of the park. Once when Joe hurried over to see what engine problem a stranger had, the man asked, "Do you play pinochle?"

We had planned to stay another month before heading to California to gather our flock and prepare for school. A policeman was checking cars coming and going to the Kodak factory, looking for out-of-state license plates. When he stopped Joe, he wrote down his license plate and gave him a warning instead of a ticket.

The state needed some out-of-state construction workers, but it also wanted the revenue it felt it deserved. Colorado's law required anyone remaining in the state longer than 30 days to get a Colorado license and license plate or pay a stiff fine. It was perfect timing for us. Within the 30-day limit, we already planned to start home.

Before we left Colorado, we called all the kids and found while we were making our plans, they had been making their own. The Snows wanted Terry to stay with them and go to school there. Terry begged Joe to allow it. An agreement was made about paying Terry's board; Terry promised he would go to school. He was very dependable.

Cathie wanted to stay in California. Skip was engaged to be married the following spring. He was staying in our house and said Cathie could stay with him and spend weekends with Jean's parents. He promised to call us if it didn't work out with Cathie. He was starting his third year as an apprentice electrician.

Janet decided her mother needed her and wanted to finish her senior year in New Mexico.

Tommy refused to go to school. Joe said, "When we get settled on another job, you'll either go to school or get a job."

With everyone except Tommy happy with their own plans, we could continue traveling this school year. I could see what real tramping was like.

Our vacation trip becomes the RV lifestyle.

There was a union hall in Tucson that had several copper mills hiring nearby. It promised enough work to last the full school year. On Thursday, Joe got a clearance to the copper mill in Oracle, 50 miles north, but he did not have to report to work until Monday.

We drove 69 miles south to Tombstone, a famous old (silver) mining town that looks much the same as it did in its heyday. For five years, it was a thriving city in the Arizona territory. Then the mines flooded, putting an abrupt end to the silver boom. Most people drifted away. Tombstone was destined to become another ghost town, like so many others, until a few hundred stubborn people came up with the idea of turning it into a tourist attraction. They nicknamed Tombstone the "town too tough to die."

To this day, it remains one of Arizona's famous tourist attractions with visitors from all around the world. It thrives on interesting legends about Billy the Kid and Wyatt Earp who walked along its boardwalk on their way to the history books. Criminals were hung in the town square where the original gallows still stands. One of Wyatt Earp's most famous gunfights took place there, and the

Bird Cage Saloon still looks like it did in the old days, when a man could buy a few drinks and a night of fun for a reasonable price. In those days, respectable ladies never walked on that side of the street.

On Saturday, we headed to Oracle to find a campground for the winter. We learned all the trailer parks within driving distance were filled with construction workers. We left the boys with the trailer on a grocery store parking lot and began searching for someplace where we could park temporarily. We hated giving up this enticing job in such a nice town.

By knocking on doors, we met interesting people, but they either didn't have room or didn't want trailers staying on their land. Many folks in those days thought people living in trailers were trashy people. Someone suggested we try the Rice ranch and gave us directions. This time Joe said, "Let's try to get acquainted before we ask."

When Goldie answered the door, Joe introduced us and said, "We're strangers in Arizona and want to learn what ranch life is like. Do you have time to talk?"

Goldie said, "Sure. We never get visitors. This is a treat. Sit down and I'll make fresh coffee."

We traded stories for over two hours until her husband, Lee, came home. He drove a truck at the mine where Joe was scheduled to work as an electrician and began telling Joe about the job.

After another hour, it was established we could stay on the hill where a mobile home had burned down. There was a septic tank there and access to water. Joe could run electric cords to the nearest power source. We agreed to move as soon as there was an empty spot at a campground.

Joe and Lee took turns driving to the mine. This gave Goldie and me use of our trucks every other day. The Rice family had a son Tim's age; the three boys had fun riding horses.

Joe and I, and sometimes the boys, helped with ranch chores, including cutting up and packaging the meat after Lee butchered a cow. That was an experience for a city girl like me.

When we mentioned checking the trailer parks again, they both shouted us down. "Seems like you might's well stay put," Lee said.

We became security guards when Lee and Goldie wanted to go somewhere. There was a question of whether their homestead rights were legal. They thought with us there that the state mine inspectors would not "come nosing around" when they were gone. Mostly, I think Goldie enjoyed having someone to chat with. Homesteading can be lonely.

Joe and Kay helping after Lee butchered a cow.

Joe offered to help a neighbor move a bee hive to a new location. Under Vince's direction, Joe fashioned a bee bonnet and taped together the bottom of his sleeves. Vince lent him a pair of thick rubber gloves while they handled the bees.

Joe picked up a big ball of bees and blew on them through his bee mask until only a few worker bees were left to protect the queen. He was proud because *he* was the one who found the queen bee. Then Vince's wife cut off her wings so she couldn't fly, and Vince placed her in the new hive.

They each carried a handful of bees to the new hive where the bees immediately surrounded the queen. Within a few minutes, all the bees had moved into the new hive with her. It was one of those serendipitous experiences that happen when you listen for opportunities. In the process, we made new friends and learned a great deal about bees.

Joe was off on weekends, so we spent many of them exploring the desert on our Yamaha bike. The problem was it didn't have enough oomph to get us both up a hill, so I did as much walking as riding. I never really learned to ride a bicycle. I did okay while riding, but to stop I had to run into something and fall off. Joe refused to teach me to drive a motorcycle until I mastered the bicycle. I never did.

On one of these trips, we found a pathway lined with rocks heading up a hill. Somebody meticulously set those rocks. It aroused our curiosity. We followed the overgrown trail for about a mile until we reached a grass-covered ridge that overlooked the towns of Oracle and San Manuel. Joe noticed every tree was circled with a rock border. "Someone did that to retain the rainfall in an attempt to keep those trees alive. There was a homestead here," Joe said.

We had explored other abandoned homesteads, but this one had no sign of a house. Only a handmade birdhouse stood as undeniable testament that someone once lived here. How can a house vanish, leaving not even a rotting timber to mark the spot? Then we stumbled upon the remains of an old mining shaft.

We had seen mining holes throughout the Catalina Mountains, but this was the first one someone attempted to fill in again. There must be the remains of a building somewhere.

Then we saw it! Not the remnants of a building, but a stone marker attached to a concrete pedestal. At the foot of the marker sat a man's boot daring the elements to destroy that small trace of human existence. Attached to the stone marker was a copper shield with words etched into it. The message, crudely scratched, was barely legible. It said a man named Earl Francis carried the makings of a house up to this ridge to erect beside his mining claim. A government rule forced him to vacate his house and claim. His answer was to sit on two boxes of dynamite and strike a match.

Our minds whirled with questions. We hurried back to tell Goldie what we found. She said, "Folks here call it 'Francis Mountain,' but that's not the official name."

Goldie gave me his sister's address. I wanted to know the whole story. His sister sent me his diary and a picture of him taken when he was 35, shortly before he died. Between his sister and Earl's diary, I learned his fascinating story.

He was a small man, five foot five inches in height and weighed barely 135 pounds. The picture shows he was almost bald and had an elfish smile above Lincolnesque chin whiskers.

He was born in Washington, D.C. After graduating high school, he held a variety of jobs but hated city congestion. He longed for the wide open spaces he read about.

In March 1953, he left Washington on a Cushman Eagle motor scooter and headed west. He carried two saddle bags with a small frying pan, sugar, salt, a few cans of food and a knife, fork, and spoon on the back of his bike. The second bag held his camera, a flashlight, hunting knife, and spare parts for his scooter. He wrapped one change of clothes in a blanket under the rack that held the saddle bags. In his pocket was $125 in cash, a diary, and a pencil to record his journey.

He was headed to California, but ran out of money in Arizona. He went to work at the crushing plant of a copper mill in San Manuel. During the five years he worked there, he spent his days off prospecting for gold along the north slope of Catalina. That's where he found his mine.

At the time Earl registered his claim, the government was encouraging prospectors to build a home near their mine and "gave" them the land to build on. It came with the stipulation that the mine must produce "enough" ore. Nobody defined enough.

It took Earl a year to build the house. He carried everything, piece by piece, up that steep trail Joe and I climbed. For a small man, this must have been challenging. His diary said the only thing he had help carrying was a refrigerator. The money for building materials came from the gold in his mine.

He mined the old-fashioned way—digging a shaft, smashing rock with a hammer and sifting the pieces to find minute particles of gold. Once a month, he took his collection to town and traded it for a few dollars needed for his house and food.

Each monthly trip was reported to the ranger. Then the government began pressuring rangers to look for freeloaders who were just homesteading on government land. By the time the Catalina district ranger discovered Earl had built a house on that deserted ridge, the house was completed. The ranger told Earl to leave.

To keep title to the land, there must be a *minimum* amount of gold extracted from a claim every month. Earl's mine did not meet the minimum even though he argued it was enough for him to live on. The ranger had no sympathy for what he believed was another illegal homesteader.

Earl's fight with the government lasted many months. He never hired a lawyer. He believed the judge would understand when he found Earl's mine produced $30 a month and that was all he needed.

He was wrong. After a year and a half of hearings, after pages and pages of reports, after hours of testimony by mining experts, the court declared his claim null and void. He appealed it.

One day he walked to the post office to get his mail and picked up an official letter stating his appeal was denied. If he did not *voluntarily* tear down his house *and return the land to its natural state*, a bulldozer would be brought in to do it.

When people who saw him in town that day spoke to him, he did not answer. They said he appeared dazed. Some thought he never heard their greeting. He walked slowly back into the desert he loved.

Later that afternoon, he and his companion—a shaggy-haired dog with the ridiculous name of Cadillac—walked to the home of a friend. He asked her to take care of his dog because he had business to attend to. She invited him in, but he didn't stay long. Later, she said he seemed upset, but he did not say anything, and she was busy preparing for company. Several hours after he left, she saw his watch and bankbook lying on a shelf.

At 5:30 p.m. as the desert was beginning to settle in for the night, a massive explosion rattled the earth. Instinctively, Earl's neighbors turned toward Francis Mountain. His American flag was flying, as usual, from the top of the mountain.

The next day the sheriff and his deputy went to investigate. They found one of Earl's boots on the ground beside the mining shaft. The only other sign of him were bits of clothing clinging to a nearby cactus. The house, only a few yards away, was undamaged unless you count the broken dish that fell from a shelf. On the wall, facing the entry, a freshly painted canvas hung crooked—like a lop-sided accusation. Earl's last painting was not one of his usual sunsets. On the canvas, he had very carefully painted: LIFE =?

In time, a bulldozer arrived and the land was back to its natural state. It's a difficult climb to the ridge. No one goes there except curious strangers like us who stumble upon it. Of the many memories during the nine months we spent on the Rice ranch, there is none that keeps popping into my mind as much as Earl's last message. What did that painting mean? I wish I had had the opportunity to talk with Earl about his philosophy of life.

Francis Mountain is still a deserted hilltop. Will it ever be used for anything as valuable as a man's life? What would the government have lost by letting Earl keep his claim? Was it as much as the cost of hours of court time and the removal of all signs of a homestead?

"It is a matter of principle," a ranger said.

With the school year over, we returned to California for Skip and Vicki's wedding. Tommy had not gone to school, and we were too isolated on the Rice ranch for him to find work. That was his excuse. However, there were things he could have helped with on the ranch. If he had been Joe, he would have found part-time work at neighboring ranches. But he wasn't Joe. During the year he traveled with us, he would rather sit in the camper doing nothing than explore with us. He had no interest in historical places and didn't care about how other people lived.

He liked riding Joe's motorcycle and the horses, but all he talked about was "going home." He said when he got back to California, he wanted to stay with Aunt Carla. His uncle agreed to take him on jobs and teach him to be a housepainter. It was the best solution we came up with. He got no more out of traveling than he had the last few years in school. He didn't like painting houses. He never married. To me, his seemed a wasted life, but it was the life he wanted. He died of cancer at age 54.

Terry was still with the Snows and begged to remain for another year. Like Tommy, he had no interest in traveling. Some of us can hear the call of the open road and others cannot. As soon as he was old enough, he joined the Marines. Except for his time in the Marine Corps, he never left the West Coast.

Joe and I wanted to continue tramping. Skip and Vickie were married and living in our house. They were newlyweds and preferred to be alone. Jean was moving to Oregon, and the family wanted Cathie to go with them, but she didn't want to live with them permanently in spite of the advantages their wealth offered.

Cathie and Tim came with us. Tim enjoyed traveling but hated changing schools where he was always "the new kid." At his age, I would have hated that, too.

Cathie looked forward to each new place and pointed out side trips. We intended to simply pass through Goshen, Indiana, but it was such an interesting place, we stayed for a month. Joe got a job close enough so he could ride his motorcycle to work, leaving the truck for us. The kids usually preferred staying in the campground with its swimming pool and the friends they made. I used the truck to go to the local library where there were many books about the Amish that fascinated me.

Most homes were built for married children to live with them and continue helping with the farm. Their houses were painted white and had no curtains or electricity. They were either talented craftsmen or farmers, using an old-fashioned plow pulled by a horse. Even the children worked from sunup to sundown unless they were going to school. Most girls went to school only long enough to read their German Bible. Boys could go to eighth grade if they wanted to. Marriages were arranged and divorce almost unheard of.

Men could not wear a beard until they were married. Their clothes had no zippers, which were considered too fancy. They were quiet-spoken, polite people who believed in hard work, discipline, and living according to their fundamentalist Bible. They were a tight community that helped each other and refused government assistance.

Because there were so many Amish in Goshen, the local culture was dominated by them. When Cathie arrived with her California bikini bathing suit, all the girls envied her. Nothing like that was sold in stores near them. I noticed that when she went to the pool to swim or lie in the sun, a number of men used the pool. A coincidence? Maybe, but I seldom saw any males at the pool if Cathie wasn't there. She was oblivious to being the main attraction.

On our last night, a group of people, including the campground managers, woke us up with a surprise going-away party. It never happened again but was a memorable sendoff for the kids.

Since we wanted to see as much as possible as we traveled, I convinced Joe that from Elkhart, Indiana, we should go through Quebec, Canada, on our way to Washington, D.C., and from there to Florida, where we planned to spend the winter near his sister and mother.

He laughed. "Is that what they teach in Yankee schools? Okay. We'll go that way."

In Lincoln, Nebraska, we stopped to see if Joe could get a short-job clearance. We got there late Friday, so the union hall was closed, but we often stayed on union hall parking lots while looking for work. We unhooked the truck in preparation for going to town to eat at a local restaurant.

As we headed to the truck, a dark ominous sky suddenly appeared. The wind picked up its pace and within seconds was blowing so hard we had trouble getting back inside the trailer. We were closing windows when whistles started blowing. Joe turned on the radio and we heard the warning.

"A tornado is headed this way. Seek safe shelter immediately."

The truck seemed safer than the trailer parked in the middle of a deserted parking lot. We ran to it and crammed in front. Joe headed sideways from where the wind was coming with the intention of circling around and getting behind it. The truck radio kept repeating to seek safe shelter. Where would we find it? The announcer assumed everyone knew.

The amazing thing was that while we franticly looked for a safe place, I saw lights in homes and restaurants. People were going about their business as if nothing was happening!

When Joe mentioned it later at the union hall, they laughed at him. "We get warnings all the time. Mostly we don't pay attention."

There were no immediate job openings for tramps so we decided to go on to Canada. We stopped at the Upper Canada Village and learned about homesteading in past times. The most fascinating was how they trained oxen to work as a team. The owner gave them directions in words. Through words and his tone of voice, these animals understood what to do. The sad part was that, once trained as a team, they would only work together. If one died, they had to destroy his partner.

Old Quebec (not to be confused with the modern city) was surrounded by high walls with only three entrance gates. The streets were narrow with houses joined together like a huge apartment complex. One street was devoted to artists painting and selling their pictures. Each evening a free performance was presented in the town square for anyone who wanted to watch.

Everyone spoke French. When we asked one man for directions, he said in perfect English, "I don't speak English." Joe got flustered and said the first thing that came to mind, "si, si, senor." Cathie doubled up laughing, while Tim tried to figure out what was funny. The man glared and turned his back.

From Quebec we visited famous landmarks in the New England states, including Plymouth Rock where we saw a replica of the Mayflower and learned 102 passengers, 25 crewmen, and enough supplies to start a new colony were all on that small sailing vessel. The ship was headed for Virginia but was blown off course. They stayed and started another colony and many people thought—and still believe—that the first colony was in Massachusetts when actually it was in Virginia.

After seeing as much as we could in New England, we went to New York. We never intended to drive through the center of that city. Our plan was to find a suburban street where we could park our trailer overnight and take the subway into the city. We don't know what happened. Suddenly we were elbowing our way toward tall skyscrapers. Joe tried to correct our mistake by turning down a side street.

A small delivery truck thought he could cut between our pickup truck and trailer. By the time he discovered we were tied together, we were both blocking traffic. Suddenly, we were engulfed in a tangle of cars and trucks like a bowl of spaghetti. Nobody could go anywhere. People leaned on their horns as if things would clear up if they made enough noise.

"I don't believe this," Cathie shouted. "Is this a dream?"

Above the honking I heard a male voice scream, "Get that thing out of here." (His language was more descriptive.)

"Do you think he means us?" I asked. Sweat was running down Joe's face. He didn't answer.

I don't know how we got into that mess, and I'll never know how we got untangled. Somehow Joe edged his way into a combination gas station and car wash.

"People are getting off work now," I assured him. "The rush hour will be over soon."

A tall, good-looking attendant with curly black hair approached. "What is that thing?" he asked.

"It's our home," Joe mumbled. "We live in it."

The attendant shook his head. "They say you can see anything in New York if you wait long enough. What's it like inside?"

"We'll show you if you'll let us park here until the traffic dies down." Joe suggested. The attendant pointed to the side of the station where a car wash was closing. "You can pull in there, but you'll have to leave before it opens in the morning."

We rewarded him with a tour of our RV and a cold beer, which Joe also needed.

His name was Camel but was pronounced Kar-mel. Although he came from Tunisia four years ago, he considered himself a genuine New Yorker. Before they finished their beer, he said, "I know a safe place to park tomorrow. If you park on a street around here, your tires will be gone and your trailer stripped when you get back."

He was off the following day and offered to show us the "real" New York. His driving was as bad as most New Yorkers. Like the rest of them, Camel constantly switched lanes, blasting his horn while cutting off other drivers and racing to be first to the next stoplight. This meant slamming on his brakes when the light turned red for crowds of pedestrians to pass. When the light turned green, the race was on again. Camel leaned out his window and yelled, "Ya wanna die?" to the stragglers.

He was right. He was now a genuine New Yorker. To my amazement, he didn't dent a fender, and no pedestrian was killed although a few leaped for the sidewalk at the last second.

Camel took us to downtown Manhattan and drove around Central Park, where he pointed out the 840-acre park's attractions, including the summer theatre where Shakespearean plays were presented free. He took us to the garment district, where workers pushed racks of clothes down the street. We rode the elevator to the observation tower on the 102nd floor of the Empire State building, and then we toured the United Nations building with a special pass he obtained from a friend.

He took us through Harlem and told us that until 1900 it was a prosperous middle-class farming village. During the 1930s, it became the intellectual, artistic, and entertainment capital of black America's famous nightclubs.

We drove through the Bowery that was once the city's liveliest theatre district. By 1880 it degenerated into a skid-row with cheap cafes, saloons, pawn shops, and flop houses. There were plans to restore it, but I don't know if it ever happened. We passed people begging for enough money to exist another day.

"What happens to them?" I asked.

"Eventually they end up in Potter's Field on Hart Island in Long Island Sound. Every week bodies are shipped there in plain pine boxes by ferry. There was a prison there (now moved to Rikers Island), and prisoners unload the boxes, dig trenches deep enough to stack caskets three high and mark them with prenumbered stones. No names." *(They called them mass graves then, but now it's communal burial plots. The prison tracks the numbers. The system started during the Civil War and is the largest taxpayer-funded cemetery in the world.)*

From the Bowery, we went to Greenwich Village, the city's bohemian center. We parked and began walking through Village streets lined with closed curio shops, discount bookstores, antique stores, and boutiques with gaudy clothes and exotic jewelry displays. We rested at a sidewalk table to have a drink and watch cross-dressers for the first time in our lives. We were fascinated by these *male* "ladies of the night."

That evening, Camel stayed for supper and played chess with Joe. It was another wonderful serendipitous experience.

The following day, we took the subway. Coming from Boston, I was familiar with them, but the kids and Joe had never used one. Joe knew there was an underground transportation system but had no idea it was so vast. He watched in amazement as people pushed through turnstiles racing for doors of cars the moment they opened. Once seated, passengers hid behind newspapers or sat mesmerized by advertisements over the heads of passengers sitting opposite them. No one talked. If two people accidently made eye contact, they quickly averted their eyes as if caught cheating.

Joe started talking to the man who sat next to him. Within a minute he had both him and his buddy engaged in conversation to the amazement of those across from us.

Joe discovered they had lived in New York most of their lives, but neither had been to the Statue of Liberty. By the time we reached their stop, Joe almost convinced the man next to him to take the day off and come with us. His buddy stood up and grabbed his arm, "Have you lost your mind? You'll get fired!"

As the door closed, Joe said, "I bet they'll never see it."

Before we left New York, we stopped to say good-bye to Camel. When we started to pull into the street, our motor suddenly quit with the nose of our pickup extended into the lane of traffic. We could not move. Joe quickly diagnosed the problem as a burnt-out coil. We were stuck there over two hours because first he had to find another coil. During that period, the traffic detoured around us as seemingly unconcerned as if this happened every day. Pedestrians either climbed over our trailer hitch or walked around the back of the trailer.

I was making a pot of coffee when I looked up to see a man's face peering in. After I greeted him, he said, "What is this thing?"

I invited him in for a closer look. While I was explaining the features, another head poked in the door. I ended up conducting one tour after another. Apparently, in 1971, a great many New Yorkers had never been inside a trailer. If either Terry or Skip had been with us, they would have been selling tour tickets.

Next we went to Washington, D.C., where there were many famous landmarks. We planned to spend several days there, so Joe called his sister to tell her we were running late but would be in Florida before school started.

She said, "You need to come as soon as you can. Mother was hit by a car, and the doctors don't know if she'll live."

Lena, now in her mid-70s, was more strong-willed than the doctors realized. Still, we packed up and hurried to Florida, knowing his sister needed Joe's support. The things we missed would have to wait until another time. We parked in his sister's yard from August till the following May.

Lena's left leg was amputated above the knee because of a blood clot from her bruises; her right leg was broken in several places and was in traction. The doctors thought she'd never walk again, but when she regained consciousness, she got out of bed, in spite of the traction, and fell. After that, we took turns sitting with her at night when she was the most disoriented. Her hallucinations worried me. Had her brain also been injured?

The accident happened when Lena was riding a bicycle back from the bar where she always cashed her social security check. As usual, she ran a red light. This was the second time she had had an accident at that same intersection. She always thought she had the right of way. She was thrown a long distance and the police were amazed she wasn't killed. Gradually, she recovered from the trauma and demanded the doctors give her an artificial leg.

Normally, at her age and with the other leg badly broken, they would not have done it, but Lena insisted Medicare would pay for it, so they ordered one. The hospital wanted her out of there as quickly as possible; she was a terrible patient, obeyed orders only when she felt like it, and her constant yelling for nurses annoyed other patients.

She finished recovering in a nursing home and learned to use the artificial leg so rapidly it amazed everyone except her. One evening, Joe and I drove Walter to the nursing home to visit her. The first thing she said, *in front of Walter,* was, "I met the nicest man. He's just down the hall. He's a World War I veteran. As soon as Walter dies, we're going to get married."

Did I forget to mention Lena was not diplomatic? All her life, she thought about what she wanted and was unconcerned if it affected anyone else. Walter didn't respond. I guess he was used to her thoughtless remarks. He outlived her and undoubtedly the veteran, too.

We entered Tim in a Florida school. He had missed a few weeks last spring because of our moving, but I believe the places he saw and the chance to actually stand on ground most children only read about in history books far outweighed any school lessons he missed.

Cathie didn't want to do her senior year in Florida. We promised she could return to California when she turned 18, so she suggested going to work to save money for her own apartment when she went back. She only lacked a couple of classes to graduate because she had built up extra credits her first three years. Her plan was to get her diploma in an adult program and then go to college. When her father died, I was notified his veteran's benefits would pay for college. Skip already had his career, and Scott had no interest in college.

Without a high school diploma and limited work experience, I wondered if Cathie could get a job. Under Joe's coaching, she got the first job she applied for as a nurse's aide in a nursing home. The charge nurse said, "You can fill out an application, but we're only hiring girls with experience."

Cathie countered, "How about this. Let me work for you for one week *without pay* while I learn. If you aren't satisfied at the end of the week, I'll leave. By then I'll have experience, and you can hire me. Either way you get a week's free labor."

The administrator, in an inner office, heard what Cathie said and yelled, "Hire that girl now!"

She worked there the rest of the time she was in Florida and developed the same love for elderly patients that I had. Later she started a nursing course.

Then marriage and traveling changed her career plans.

We were in Florida almost two months before I heard about Reverend Lin Wilcox and his wife, Artist. For 26 years, Lin was a pastor at a Baptist church in Pennsylvania. In 1965, they went to Florida on vacation and discovered "The Range Line"—a narrow strip of land along Hwy 441 between Palm Beach and Pompano Beach.

To the east lies the coast with sandy white beaches, high-rise hotels, and curved palm trees bending forward to welcome wealthy tourists who flock there each winter. To the south lie the Everglades, a wilderness of saw grass inhabited by rare birds and wildlife.

The Range Line itself is fertile farmlands, interrupted occasionally by a field where white egrets reside on the backs of grazing cows. Above it, whipped-cream clouds add to the illusion of harmony.

On back roads where tourists never go, migrant farm workers live while they pick crops for that farmer. Their houses are tarpaper shacks, wooden barracks, or dilapidated shanties. A family often crowds together on one big bed with a lumpy, bug-infested mattress or on cots with a blanket folded for a mattress. They have limited electricity but no refrigeration. Usually, there was a wash basin sink with a cold water faucet in each house. Cold water showers and toilets were generally in a community building.

Most houses had broken windows with the screens torn off. Some windows were covered with cardboard to keep out the mosquitoes and flies that covered trash piles outside. Opening the door provided the only ventilation in Florida's summer heat and humidity. When it rained, the roof probably leaked.

I decided the leaking roof summarized the farmer-migrant relationship. Because he provides a free place to live, the farmer feels the migrant should fix any leaks and put screens on the windows that are broken; the migrants complain about conditions but see no reason to fix a wealthy farmer's roof or windows.

To be fair to the farmer, this was not always the case. When the houses were new, they had glass windows and screens on windows and doors. Some farmers tried to keep the houses clean and comfortable with chairs, tables, and such. However, by the end of the season, when that group of migrants headed north, the furniture and windows were broken, screens torn, and trash strewn everywhere.

One farmer said, "I stopped trying. These people don't take care of anything."

It was understandable. When everything belongs to somebody else and you think that person is cheating you, would you care what happens to *his* house? If things get bad enough and the farmer can no longer make a decent living, he can sell his land; the migrant is trapped in the system because he has neither the money nor the education to get out. Some may dream of that "other world" where everybody has a blanket on their own bed and a new pair of shoes, but even children know their lifestyle is a trap from which very few escape. All the whites, blacks, Puerto-Ricans, and Mexicans I talked with seemed to be almost philosophical about their lot in life. They had grown up in the migrant system.

Lin said things were better now than in 1966 when he and his wife decided he was needed more here than in an established church. They moved to Florida to bring God to the migrants who desperately needed something to give them hope. In 1971 when I arrived, Lin was 44 years old, a little on the heavy side, and nearly bald. He and Artist were the only ones I met that cared about Florida migrants.

With his savings, Lin bought a bus to take children to the beach and ball games—places they didn't know existed. He went to local businessmen and charity groups begging for donations of food, clothes, dishes, and blankets. Gradually, his small mission grew into a nonprofit corporation called Operation Concern.

When I heard about his mission, he was in need of volunteer nurses to go into the camps and see if there were sick people that should be in a hospital or needed medicine. He found a few doctors willing to see patients who could not pay, and volunteers took them there.

Three days a week I went to Lin's headquarters and loaded my pickup with whatever was available—sometimes clothes, sometimes food. My real objective was to check on the health of those in the camps he assigned to me.

Complaints were from an older sister babysitting while her parents worked in the fields. "My sister is all over sores. Can you fix her?"

Running sores on bare legs and feet were common; all I had was alcohol to clean the sores and calamine lotion to alleviate the itching. Another common complaint was lingering coughs and belly aches. Yet it wasn't always that simple. One day when I was distributing hot soup, a woman tugged on my arm. "My Juan's been swallowin' up. He looks funny. Can you come?"

The minute I entered the dark, stuffy room, I knew it was too late. Still I went through the motions while I tried to think how to tell her the baby was dead.

Death was no stranger in the Range Line. Bodies were often buried in shallow, man-dug holes or were found floating in the canals. When there is not enough money to buy food for the living, how can you afford a funeral for the dead?

One day Lin asked me to go to a camp that was not on my usual route. As I pulled in front of a long row of dilapidated wooden buildings, it seemed as if a million flies swarmed around the trash strewn by them. A black girl who appeared to be 14 or 15 stood in the nearest doorway. Her hands supported a pregnant belly while a half-naked toddler clung to one of her bare legs. Half a dozen puppies wrestled on the cluttered floor behind her.

"I'm looking for Jed Brown," I said, stepping over a lame cat with a new litter of multi-colored kittens. "Do you know where he lives?"

"He's my daddy. Ain't here. He's went to the field."

"He's working? They said he was sick. I have medicine for him."

She nodded solemnly. "He needs it. Water here tore his belly up awful bad."

According to the report I'd been given, her father had stomach ulcers that hemorrhaged the day before. I was to either take him to a hospital or, if he wouldn't go, give him a bottle of Maalox. I handed the girl the Maalox from my shoulder bag. "Do you need any food?"

Her hands were closing on the bottle when I saw a look of fear on her child-mother face. Snatching up the toddler by its arm, she disappeared into the shadows of the room behind her.

"What in hell you doing here?" thundered an angry voice behind me. Turning, I found myself looking into eyes filled with hate. Before I could answer, he said, "I told Wilcox I don't want you people comin' here no more! Can't you read?"

He gestured with his shotgun to a huge sign I had not noticed before.

KEEP OUT!
TRESPASSERS WILL BE SHOT

"Think I don't mean it?" A sign on his shirt announced he was camp manager.

I tried to ignore the gun. "Mr. Brown is sick. I brought him medicine he can't afford."

The manager spat at my feet. "That's 'cause he spends it on wine." He didn't mention that *he* bought the wine and sold it to Mr. Brown and the others for more than twice the price he paid.

"Since I'm already here, please let me leave food for the children," I pleaded. "Surely you don't object to our feeding the children."

"The hell I don't! Long as you give 'em handouts, they ain't gonna work. They're a no-good bunch and nothin's gonna change 'em. Now get outta here. You come back, I swear to God I'll shoot you for trespassing."

Another migrant I will never forget was a proud old black man who was 73 when I met him. In an entire lifetime of following the migrant work cycle, he never saved any money. He was still working whenever he could with arthritis so severe he crawled on his knees, picking the beans close to the ground—the beans left after more capable pickers had stripped the stalks.

When it rained, his arthritis was often so bad he could not crawl. At those times, I would sit with him on the stoop of the room where he lived alone and listen to stories about the family he once had. It seemed that listening to him meant more than the food I brought. I wondered if he was an outcast of the camp.

One rainy day I stopped to see him on my way home, even though it was not on that day's assignment. It was raining so hard that many migrants weren't working. The rooms were always stuffy on days like this, so all the doors were open except Grandpa's.

I suspected something was wrong. When no one answered my knock, I opened the door and saw his form curled up on a cot in the windowless room. The only light came from a single low-watt bulb dangling from the ceiling. After brushing a cockroach off the cot, I placed my hand on his forehead. He didn't seem to have a fever. "What's wrong, Grandpa? Are you sick?"

"It's the misery," he answered trying to straighten his legs.

"Can I fix you something to eat?"

"I'd be proud was you to open a can of beans."

I pushed aside some dirty dishes and set a can of pork and beans on the bench that served as both seat and table. I was looking for a cooking pot to heat the beans when I realized there was no stove. "Don't you have a stove?" I asked.

He closed his eyes as if in shame. "It cost extra."

"Why didn't you tell us? Reverend Wilcox would have given you a hot plate."

"You folks been good to me," he replied. "Ain't right I should ask for more."

His attitude was typical. I found Range Line people accepted whatever I brought, but never asked for anything. That was why I continued bringing candy to the children. I knew they were better off without it, but it seemed to mean so much.

I started the practice when I had an entire bag of candy left over after Halloween. When I brought it, I saw smiles on faces that I had never seen smile before. Candy was something these children never got. It was a small thing I could give them from my own pocket. The project provided important things.

Yet, who can say what is important? Maybe to eight-year-old Susie the most important gift she ever got was a pair of brand-new, never-been-worn-before patent leather shoes I bought her when I discovered she and her older sister took turns going to school. They shared one pair of sneakers that were too big for Susie until she stuffed the toes with paper.

Buying sneakers would be practical. I remembered as a child wishing for black patent leather shoes; they weren't practical then either. I wanted her to have one memory to cherish that wasn't practical. I still remember her excitement when her skinny little arms hugged them.

Florida holds different memories for all of us. Joe, Cathie, and Tim's memories are not the same as mine. Even my memories of this Florida clashed with my past life in Florida. It is as if they were two different places.

Most people think of Florida as white sandy beaches and canals where black people sit holding cane poles to tempt a catfish, and turtles sun themselves on rocks while a glassy-eyed alligator glides through algae-covered water. That is the tourist-book side of Florida. They never picture the devastation after hurricanes. (I went through one hurricane in Florida when I lived with Jim, and it is something you don't forget.)

Now when I think of Florida, it is Operation Concern where I saw babies lying in boxes in the field while parents spent their lives picking vegetables for someone else to eat. Excited voices calling, "Here comes the Candy Lady!" still echo in my mind.

I have volunteered at other jobs over our travel years, but none inspired me like helping these migrants, and none impressed me more about the importance of education. Reverend Wilcox said, "We give them handouts when we can, but handouts are not the answer. We must teach them how to help themselves."

Reverend Wilcox borrowed money and purchased seven acres of land on which there was an old grocery store and a bar. He turned the grocery store into a library and learning center where children and adults received tutoring in the skills they lacked in reading, writing, and math. Then he turned the bar into a chapel that doubles as a youth assembly room as well as a gathering place for activities. The bar counter was cut down to table size for senior citizens and children who come for one hot meal a day. It is the only meal some will have all day. Yet, of all the things Lin gave them, the most important was *hope of a better tomorrow*.

Cathie had flown back to California. Tim's school year ended in May. It was time to head west to check on the other kids and start to Alaska. We had been living on the road for two years and both Joe and I loved the traveling lifestyle. Joe learned something new on every job. This was the happiest he had ever been. Learning meant as much to him as it had to my dad. Neither got the education they wanted, even though the reasons were different.

When we returned to California, we put our house up for sale. Skip, his wife and their first son were living in it but planned to move to Oregon when Skip's apprenticeship was finished. At the time, it didn't make sense to pay for a house we thought we'd never live in. If only we knew what the future holds, how many decisions we would make differently!

We saved money working on the road but wanted a bigger Airstream, even though it seemed as if Tim was the only one who would continue traveling with us. We traded our old Airstream for a new 30-foot Airstream. The house market was starting its boom years, so the house sold quickly for our asking price. We sold the house for twice what we paid for it, enabling us to pay cash for the new Airstream and gave us the start of a financial cushion we had lacked.

The road to more adventures stretched ahead with the promise of a summer spent in Alaska.

CHAPTER 12

Alaska Adventure (1972; 1974; 2003)

OUR ADVENTURE BEGAN when we crossed the Canadian border. We knew we weren't supposed to take guns, but all the years we traveled, Joe kept a revolver hidden in our trailer (our *home*). Since we boondocked a lot, it made him feel safer. We loaded our camper with groceries knowing prices in Alaska were higher. Just as we approached the border, I remembered I had a dozen postcards with U.S. postage on them. Joe said, "We'll go back to town and find a mailbox."

We made a U-turn and headed back. Apparently, that came under a suspicious act to the border patrol. When we got back to the crossing 15 minutes later, they pulled us aside. A guard began at the back bedroom opening drawers and closets and tossing the contents on the bed. With a total mess strewn behind him, he reached the kitchen pantry.

A smaller two-shelf cabinet with books in it was above it, and the gun was behind our books. He started with the pantry, pulling everything off the top shelf and piling the groceries helter-skelter on the stovetop. As he worked his way down, he dumped things from one shelf to the empty one above or tossed it on the stovetop and sink.

Then he opened the book cabinet. He started taking down the books one by one, but there was no place left to put them, so he was shifting them to his left arm. In a matter of seconds, he would get to the book the gun was behind. He looked around for someplace to dump the books when Joe distracted him. "Do you mind telling me what you're looking for?"

The guard was tired of holding the books. He started to pile them in the empty space. "You tell me why you made a U-turn at the border and then came back 15 minutes later. What were you hiding?"

"If you'd asked me that, it would have saved you a lot of time and work."

Joe explained about the postage. Our excuse seemed dumb, but the guard believed it. However, he still wanted to look in our camper. When he saw it was loaded with groceries, he seemed relieved to have found something we were doing wrong. "You can't take all those groceries to Canada. It isn't allowed."

Joe said, "We aren't going to Canada. We're just passing through on our way to Alaska. I'll be working in Alaska all summer."

He checked with his superior and reluctantly let us go with our trailer in total disarray. After that, whenever we crossed into Canada, we left our gun with friends.

The old Alcan Highway began in Dawson Creek, Canada, and today is officially called The Alaska Highway. It is much different from the mud and gravel road we traveled in the summer of 1972. The highway was built by the U.S. Government as a defense and supply route for installations in Alaska after the Japanese took over some of the Aleutian Islands during World War II. The U.S. Army Engineers cut through nearly 12,000 miles of wilderness—places where only dog sleds had ventured before—and bridged more than 100 rivers and streams.

This awesome task was a joint project between Canada and the U.S. In only six months they had a road passable for jeeps and trucks. Throughout that winter, both armies worked on it in 40-below-zero weather. When summer came, the work continued, even after rain turned the road into a jelly-like quagmire. Thirty years later, it was said to be one of the finest gravel highways in the world. Today the road is paved making traveling easier and safer.

Protecting our Airstream from flying gravel and rocks.

Our Ford pickup and Alaskan camper were towing a new Airstream, so we made preparations to protect the trailer from rock damage. This included a bug screen on the front of our pickup truck to prevent bugs from blocking the radiator and deflect flying gravel from hitting the windshield. We used plastic covers to protect our headlights because a broken headlight could prevent an oncoming vehicle from seeing you through the thick dust. We put mudguard flaps on the rear truck wheels to prevent a constant barrage of tiny rocks from literally stoning the front of our trailer. To further protect it, we attached flattened cardboard boxes over the leading corners and windows where flying rocks would cause the most damage.

These preparations were made at Milepost 0 in Dawson Creek, British Columbia, where the Alcan Highway began. Driving the Alcan made me more aware of the forces of nature. Here survival can depend on the help given by a total stranger. In those days, it was against the law for anyone to ignore someone waving for help. Those who travel this road share a special kinship. Even with today's communication systems, people do not drive past a disabled vehicle that is asking for help.

When the spring shackle between two of our trailer wheels broke in half, Joe removed one of the trailer wheels and chained the axle up. While he did this, a number of cars stopped to ask if we needed help. One of them was Bud and Marilyn Barber's, whom we had first met near Las Vegas.

At that time, we were parked on the side of the road because our engine was overheating. Bud stopped to see if we needed help. He and Joe found they were both navy veterans as well as electricians. They were headed to Alaska also, but we never expected to meet them again, especially under similar circumstances. It was the beginning of a friendship that lasted over 35 years until both Joe and Marilyn died.

Joe had everything under control, so we told them we'd watch for them as we traveled. Pulling the trailer on three wheels, we stopped at Seagull Creek where there were two gas stations; neither had a replacement part. It would be a couple of weeks before a new part could be sent in from the States. One mechanic suggested a welding shop at Teslin, so we limped there on three wheels. The only welding shop had a sign on the door.

Gone to a turkey shoot
Come back tomorrow

We settled down to wait. Shortly afterwards, we saw an Indian lad walking around our trailer. Joe went outside to see if he was up to mischief. He was a typical teen-ager interested in our motorcycle.

He told Joe, "I know a lot about motorcycles, but I don't have one."

Joe recognized the longing in his eyes and offered to let him ride ours for one hour. He took off with the biggest smile I've ever seen, promising to be back in an hour, and headed straight for his village—wherever that was. We had some anxious moments while he was gone, but even if we lost the motorcycle, I was glad Joe did that. There was no guarantee we'd ever see him again, and yet there is something about being on the Alcan that makes you trust and care about each other.

The boy was back exactly one hour later and wanted to pay Joe for letting him use the bike. When Joe refused his money, it upset him. "I must give you something! Our custom demands it."

That evening, he returned carrying a large paper sack filled with dried moose meat. I looked dubiously into the sack. "How do you cook it?"

"We eat it like it is, but you can boil it in water and dip it in grease."

"Grease?" I echoed. "You mean dip it in butter?"

"No, ma'am. Don't you have grease left over from cooking? Makes it taste good."

That night we tried eating it "as is." I started with a small bite, but the more I chewed it, the bigger it got. I took half of it out of my mouth and started again. I never got beyond that first bite.

The next day, when the frontiersman arrived, he began working on our shackle. It took him all morning because he kept stopping to tell Joe bear-hunting stories, some of which sounded as if they had been doctored with a good dose of imagination.

I boiled some of the moose meat for our lunch. I did not have grease, so we dipped it in butter. Perhaps that was why it was tough and tasteless. Or maybe we were not hungry. When Joe mentioned it to the frontiersman, he suggested I cook it with vegetables like a stew.

After lunch, we went a short distance until we saw a sign:

AUTHENTIC INDIAN HANDCRAFT

There was no one around, but the cabin door was ajar. As we started toward it, a pack of dogs of varied and uncertain ancestry came yapping to greet us.

"Who's there?" called a voice.

"We came to look at the handcraft you're selling," Joe called. I saw that the visible part of the cabin was in shambles. After a few minutes, a white man shuffled out, hitching up suspender straps over the top of long-john underwear. Based on color and odor, I suspected he had worn it a long time.

"My woman and me slept in," he apologized, rubbing his whiskered face. "Had a turkey shoot over to Teslin yesterday, and we was celebratin' all night."

"Then you won," I said. "Congratulations!"

A broken-toothed smile lit his face. "Don't got to win to celebrate." He turned his head and called over his shoulder, "Come out, woman." He said, "She's a full-blood," as if daring us to object.

An ageless, raw-boned, Indian woman appeared carrying a cloth bundle that she spread on a wooden table in front of a rope stretched between two trees. Half a dozen varied animal skins hung from the rope. While I examined the items, the trader kept up a steady conversation. He told us he was an American who came to the Yukon 25 years before. He thought it very funny when Joe referred to the Indian as his wife.

"She's my woman." he emphasized. It seems that title allowed her to clean his cabin, wash his clothes, and cook his meals. I hoped she was a better cook than laundress and housekeeper.

"I'm a trapper," he announced proudly as he pointed to a line of curing animal skins. "See that fox hide? Shot him right from my doorway. Sent my woman to go get it, skin it, and cure the hide."

I selected a few pieces of jewelry to send home to our children, not because I thought them worth the small asking price, but because I liked this whiskered old man and his woman who never spoke except to quote a price. Any other question she answered with a nod, a shrug, or a frozen-faced stare. Yet, through all the bartering, there was a glint in her eyes that made me think she was silently laughing at all of us.

Even if she had wanted to speak, there wasn't much chance. The trader talked non-stop. He was reluctant to let us leave, saying he didn't have many visitors. "Even in summer, tourists go zipping past," he said. "Don't know where they're in such a hurry to get to."

Finally we broke away and started to our truck when he came running with out-stretched arm and a rabbit foot dangling from his hand. "I didn't buy that," I said.

"It's for the boy," he replied. "I'm givin' it to him just for the hell of it."

That night, I made a pot of moose meat stew. We ate it with salad and bread. The salad, bread, and vegetables disappeared, but we ended up with an almost full pot of moose meat and liquid. We still had three-fourths of dried moose meat in the sack.

I hate to throw food away. Joe came up with the idea of mailing the dried meat to our friends in California to serve at their bridge party. Later they said everyone had fun trying to chew it, but if we had any more, to keep it for ourselves.

Where else but Alaska can you stand on a suspension bridge and watch an ice-clogged river's spectacular display when the ice breakup comes? Watching those slow-moving chunks of ice, I thought again about the violence hidden behind nature even when she wears a calm face.

Collection is at Milepost 635 at Watson Lake.

At Milepost 635, Watson Lake has what is undoubtedly the largest display of signs anywhere in the world. It started during those lonely days when men were cutting the road through the wilderness. A homesick soldier put up a sign with the name of his hometown written on it. The idea caught on. Other soldiers put up signs from their hometowns. When the public began traveling the road, they did the same.

In 1972, it was an interesting collection, but when we drove the Alaskan Highway 20 years later, the collection of signs had grown to thousands and has become a famous landmark.

At Milepost 974, I was intrigued with Indian cemetery "spirit houses." These miniature houses stand over the graves instead of stone markers. All have windows, many with curtains to give the spirit more privacy. None had doors. When I peeked into some of the houses, I discovered there was a table or chairs. In one I saw an old trunk that I suppose contained the spirit's most prized possessions.

Haines Junction at Milepost 1016 is where the Alcan turns northward into the foothills of the St. Elias Mountains. The Haines Highway begins here and runs for 159 miles southeast to the ferry port of Haines. From Haines, you can take a boat, operated by the Alaska Ferry System, to the island ports.

However, we stayed on the Alcan until Milepost 1221, the Alaska border. We had traveled over 1,000 miles on dirt roads with the exception of 15 miles of concrete going through Whitehorse. Today, that road is called The Alaska Highway and is paved all the way. So unless you drove it while it was still a narrow, rutty, dusty bedrock gravel road, you cannot appreciate what a marvelous invention pavement is.

When we reached the Alaska border, we stopped to take a picture. While Joe was changing one flat tire, Tim went to the other side and discovered another one. We had eight flats going up and five coming back. The next time we went, we had Michelin tires and no flats.

Dad! This one is really flat.

When we reached pavement after days of driving 35 miles an hour, Joe was tempted to speed up, but soon discovered this pavement was not like the superhighways in "the Lower 48." Because it is constantly at the mercy of freezing snow and ice, the pavement buckles and cracks, making travel almost as slow as it had been on the bedrock gravel. Still, the absence of dust and flying rocks was a welcome relief.

The real challenge of the Alcan was behind us, and, in spite of the flat tires, we felt a sense of accomplishment that only those who have "come through" can understand.

We did not come through unscarred. The bolt shook loose on the seat of the motorcycle and the seat bounced off and was lost somewhere in the cloud of dust that followed us. One of our closet doors would not open, and when it finally did, we couldn't get it closed again. At one point, our outside storage compartment flew open and vomited our sewer hoses over the highway to become lost in dust. So many screws came loose that tightening things became as much a part of life as brushing our teeth at night.

What was it like? Eleven hundred miles of narrow dirt road with dust so thick you can't see; rocks flying at you like bullets; and clouds of mosquitoes waiting for you to open the door. It was also forests, snow-capped mountains, and lakes that reflected the sky while fish jump up daring you to throw in a line. For me, it was a road of memories of a sack of dried moose meat, an old rabbit's foot, and the feeling of kinship with all kinds of people.

Entering from the Alcan as we did, it was difficult to comprehend the magnitude of a state covering 586,412 square miles and four time zones. It was easier to understand when I visualized picking Alaska up, squeezing it together, and then laying it on top of a map of the United States. It would have covered all of Texas, New Mexico, Colorado, Kansas, Arkansas, and part of Louisiana.

I wanted to meet the Lentz family that Joe had stayed with during the forties, so we made a side trip to the Matanuska Valley. We found Ma Lentz at the Pioneer Home in Palmer. Her husband, Joseph, had long since died. *(Ma died at the Pioneer Hospital in 1975. Four of her children spent their entire lives in Alaska.)*

Ma said, "We didn't know what we would find in Alaska but rumors were frightening. People said the snow got so deep it would bury a house and the mosquitoes so big you could shoot them with a gun. The trip to Alaska was awful. The train to Seattle was crammed with crying babies and scared kids; parents were exhausted. We thought we'd get some rest when we got on the ship, but it was just as bad. Worse really, because now most of us were seasick.

We kept telling each other things would be better when we reached Alaska, but when we got there, it wasn't any better."

It seems that the workmen, who were supposed to have a temporary tent city ready for the colonists to use while they built permanent homes, arrived *after* the first boatload of colonists. The workers were kids and young men from the CCC (Civilian Conservation Core) program. Most had no building experience. On the same ship as the workers were 1500 tons of farm machinery, tools, food, and the colonists' household goods.

This first group had to stay on the ship four more days until they had a canvas tent city ready. Each family was given a single washbasin and a ration of food including "boat" eggs. (In those days, eggs were shipped in poorly refrigerated vessels, so by the time they reached Alaska, they had a distinct odor and taste. Some people never got used to it.) Until their own household goods arrived, they boiled coffee in a tin can and cooked their food in the same washbasin they had to eat from.

They didn't know where their land was until the second ship load of colonists arrived and a lottery to draw for their land was held. Families were allowed to exchange their tracts with others. This meant those who had become friends were able to get farms in the same section. However, whether their neighbors were friends or strangers, they knew they had to help each other.

Ma said, "All summer we lived in the tents and fought the mosquitoes. There is no privacy when tents are side by side. We heard our neighbors arguing and children crying day and night. We prayed things would get better, but they kept getting worse. It started raining and didn't stop for days. It was cold and wet and mud was everywhere. Everybody got sick. Measles and colds spread from one tent to another. There was continuous coughing, sick people moaning, and children crying."

Many thought the government had deliberately lied to them. When people are unhappy and have nothing else to do, they complain. In tent city nobody was happy, and they could do little except wait impatiently for their houses to be built.

"Some of us wanted to build our own houses so we knew they'd be built right, but they wouldn't let us. A lot of folks just gave up and demanded to be taken home," Ma said. "We didn't have anything to go back to. We had to make do. We made friends with the strangers who shared our misery. Might seem peculiar, but we became closer to strangers than we had ever been to neighbors we'd known for years."

(Within four years, 60 percent of the original colonists were back in the Lower 48.)

Although the days are long in Alaska summers, the building went slow. When time was running out, the colonists were finally allowed to finish their houses themselves.

Ma and Pa Lentz, like others who stayed, eventually paid off their debt to the government.

Of the many stories Ma Lentz told me, one comment summarized them all: "I found out that things I used to think were important didn't matter at all. I would gladly swap my good china dishes for a hen that laid eggs."

Our experiences teach many lessons. Understanding their helplessness against nature united the colonists. I found that same cohesion still existed throughout Alaska. We were welcomed as family every place we visited.

When we reached Anchorage, we settled in a trailer park to get our rig clean, do our laundry, empty two holding tanks and fill the third with fresh water. Joe had "signed the book" and was waiting for a job clearance. One day, Lynn Rogers drove through the park on his way home from work. He wanted to travel but had two little girls and thought it was impossible. The closest he could get to RVing was camping on weekends.

Our trailer caught his eye. "I've seen that trailer somewhere before," he thought. In looking it over he noted our logo on the side said, "Peterson's Roots." He rushed home and picked up the latest copy of his RV magazine. Thumbing through, he saw another Airstream with the words "Peterson's Roots," but it was an older model and shorter. He was holding my first published article, with a picture of Joe and me sitting in front with our membership number above our heads and the words "Peterson's Roots" clearly visible on the side.

He was so excited he drove back to the trailer park and insisted we come to his home for dinner the next evening and meet his family. This man was obviously suffering from "hitch-up fever."

When we arrived the next evening, he could hardly contain his eagerness, but his wife did not share his enthusiasm. Her questions indicated she was afraid of living full-time in an RV. After we left, they stayed up all night until Lynn's excitement overwhelmed her resistance.

The day after our dinner with them, Joe got clearance for a job in Juneau. We must leave immediately, but we would remain in touch with Lynn. There are no roads in or out of Juneau, which is in the southeast panhandle of Alaska. It is isolated by mountains from Canada, which lie to its east, and water and impassable ice fields block access from other directions. The easiest way in and out is by air, as Joe discovered after the war. The only way to get a truck and trailer in and out is on the ferry.

When we arrived in Juneau, it was late, so we parked overnight by Mendenhall Glacier, where other RVers were "boondocking" (parking without hookups). The next morning, I left Tim with the trailer while I drove Joe to work. Joe told the men on his crew I was on the way to find us a trailer space.

Johnny said. "She won't find any."

"The phone book lists three parks," Joe replied.

"They're full. I know. I've been looking for over a month. Andy has, too. I'm still in the parking lot down by the wharf. Your wife's wasting her time."

"You want to bet on it?" Joe said.

Johnny frowned. "You know something we don't?"

"No, but I know my wife. She'll find a place," Joe said with more confidence than he felt.

Johnny and Andy were right when they said all the parks were full. That's the information I received when I called. It was easy to say no on the phone. Sometimes you need to look people in the eye. I drove past a big "NO VACANCY" sign posted at the driveway of the first park and went to the manager's office. As soon as I stepped inside, I sensed she wasn't friendly, but I was already there. I asked if she knew of any place we could park a 30-foot trailer. "Nope. No place in Juneau. Some folks park out by the glacier."

"They don't have hookups," I said.

She nodded. "Folks are gettin' by, but don't drink the water. Tastes icky."

At the second park I saw several vacant spaces between mobile homes in spite of the fact that a huge sign stated: *"NO VACANCY."* There was no sign indicating the manager, so I knocked on the first door. A young woman with a child in her arms pointed, "Over there. If you're looking for a space, forget it. They're closing this park down. They said we got 60 days to get out, but there's no place for trailers in Juneau."

This information started an idea buzzing. People didn't seem to know the difference between RV trailers and what we call mobile home trailers.

The manager was an older woman with a nice smile. "I saw your sign, but I see there are several vacant spaces. I'm looking for a space for a *travel* trailer—not one like those. We only need it for a couple of months. Can you rent us a space for 60 days?"

"Sorry. Owner says don't take no more trailers."

I told her Joe would be working here probably until September. "It is so hard to find a space for *a short time*. Most parks want permanent people. We travel most of the year." With her curiosity aroused, she invited me to have a cup of tea. As we passed through the living room, I noticed many crocheted doilies on arms of chairs. I picked one that looked different. "I've never seen a doily like this. Did you design it yourself?"

Her eyes sparkled with pride. She hadn't designed it but was proud of it. She talked more about crocheting while she made the tea.

"Do you do any other crafts besides crocheting?"

We were soon involved in an animated conversation that I smoothly slid into a discussion of children and grandchildren. She brought me pictures of hers, and I told her about Tim and how it was hard for him to change schools so often. She was amazed to hear about job-hopping with children and wanted to know more about where we had been.

We were having cookies with the third cup of tea when she confided the owner planned to build an apartment complex on this site when they get the trailers out. Someone was supposed to build a big trailer park out by the glacier, but they had trouble with permits, and now a late spring thaw further delayed the project. "Don't see as how they'll have it ready by fall," she admitted.

"Will the owner build the apartments this fall?"

She sighed. "He's wantin' to soon's he gets them trailers out. They got nowhere to go 'til they build the park."

"I have an idea," I said. "I have a *travel* trailer. It isn't like these. We can move out of a park in 15 minutes."

"Gosh! I wish we could."

I forced eye contact. "The owner is losing money while he waits for these big trailers to move. He could rent the empty spaces to *travelers* with *small* trailers like mine. All *travel* trailers can be gone within 24 hours."

She seemed excited and agreed to call the owner. I told her we'd come back after Joe got through working. "We'll bring our *travel trailer* so you can see how easy it is to move. I'd love living next to you. Maybe you can teach me more about crocheting."

I was almost positive that we could talk the owner into renting to construction workers. There were so few RVs in Alaska at that time. Once we showed him an RV trailer, I thought any smart owner would understand he could make money while he waited. However, nothing is a sure thing, so I decided to look at the remaining park on Douglas Island.

I tried a different approach when I saw this park already had both mobile homes and travel trailers, and I couldn't see any empty spaces. I had little hope, but it was a chance to practice my persuasion skills.

"My husband is working here for the summer, and we've been unable to find a place to park our Airstream *travel trailer*. Looks like that one," I said pointing, "so it doesn't take much space. This is such a pretty park. Our son would love it here."

"How old is he?"

"He's 12. We travel from job to job. We'll only be in Juneau until fall. We need to be settled in the States when school starts."

Her curiosity about living full-time in our RV was obvious. She burst out, "That's what we want to do when my husband retires. I could let you park in a space we were saving for the motor home we plan to buy next year. There's no electricity, but you can run an extension cord to an outlet."

"That's no problem. My husband is an electrician. Maybe he can install an electric outlet. Then you can use it for your motor home."

That evening the three of us drove to both parks. We all liked the one on Douglas Island best because it was on the waterfront. We settled in with temporary electric until Joe had time to install a permanent setup.

The next day I again drove Joe to work so I could use the truck. Johnny asked sarcastically, "Well, did your wife find a place to park?"

Joe knew the whole crew was expecting to hear a tale of woe. "Actually, she found *a bunch of* spaces."

They thought he was joking and started laughing.

Joe continued with a bit of a gloat. "We decided to take the one on Douglas Island, but I talked to the owner of the park in town. He said he'll rent spaces to anyone in a travel trailer if you promise to move out when he's ready to build an apartment complex. Probably won't start till next spring."

(A few days before we left Juneau the end of September, we passed the park. A dozen RVs were parked along with several mobile homes. The apartment complex was rescheduled to start next year. We had helped the park owner as well as Joe's working buddies.)

During the summer, I decided to get a job since Joe was working long hours. I didn't have a license for Alaska, but I had always wanted to work in a doctor's office where I could work under the doctor's license. Nobody wants to hire someone for a short time—unless they need temporary help. At the first office, I asked if they wanted someone to do vacation relief for their nurse. (Qualified vacation relief is hard to find everywhere.) The doctors ended up passing me from one of their friends to another, so that I worked most of the summer.

Joe heard the Juneau assistant district attorney was looking for a chess partner. He found Bill and they began playing on a regular basis, although usually Joe won. Bill was entering the annual chess tournament and suggested Joe enter, too.

In the initial rounds, Joe didn't lose any games. However, this was a double-elimination contest, which meant you had to lose two games before you were out. In the first round, Joe beat Bill, but except for that loss, Bill won all his other games. This put them both in the semi-finals.

The semi-finals took place in the school with a classroom full of onlookers. Again, Joe and Bill played. This time it was two out of three games. Joe easily beat Bill the first game, but the stress of playing in front of a group of people got to him. He got rattled and made a bad move in the second game and lost. When it came to the third tie-breaker game, he was a nervous wreck. He said he played like an amateur. Being an attorney, Bill was used to pressure and being in front of people; it gave him a big advantage on this game. The result was that someone who had not lost any games came in first and Bill came in second. Joe came in third, but there were only two trophies.

Soon it was late September. The snow line had crept down the side of the mountain visible from our trailer. "Better get out before the Taku winds come," our landlady warned.

We had taken the camper off the pickup during our stay there, so we planned to spend the last morning getting it back on the truck, loading everything up, disconnecting sewer and water hoses, and unplugging the electric. We already had a ferry reservation to Ketchikan, where we planned to get off and drive the rest of the way to Seattle. At that time of year, both ferry lines—one to Ketchikan and the other to Seattle—ran only once a week.

While Joe worked his last day, a road crew put new planking on the bridge to Douglas Island. Unaware of how slippery new planking can be when wet, and being in a hurry to get home, Joe skidded on the bridge. He managed to get off the bridge before his truck turned upside down in a bar ditch. Had he gone off the side of the bridge, he would surely have been killed. If he had had his seat belt on, as he was supposed to, the top of the truck would have crushed his head. The top of the truck was lying on the dashboard after the truck landed, but Joe had been thrown on the floor.

Again he beat the odds. He escaped with a few cuts and very sore muscles. Our tow vehicle did not fare as well. Thinking it was beyond repair, the police had it towed to a junkyard. This left us with no way to get our trailer or camper off the island. All I could think about was what would I do if I lost Joe?

Joe was trying to find a solution. He inquired about buying a new truck. No trucks in Juneau were equipped for towing our size trailer. To have either the parts to repair the old Ford, or a new truck shipped in, would take from two to six months.

That night, while Joe tried to decide what to do, the first Taku of the season hit with strong, gusty winds and cold rain. Porches were blown off, trees fell, and power lines were down. The next morning, when the radio station was back in operation, the announcer said, "We had some brisk winds last night. This is our annual warning to get your shingles nailed down before winter comes."

Before winter comes? The temperature had bottomed out at 30 degrees, and the wind-driven rain cut to the bone.

In the meantime, Joe had a plan. He called a work buddy, and the two of them went to the junkyard to see how badly the truck was damaged. The most obvious problem was the smashed windshield, the top of which was touching the dashboard.

Joe and his buddy managed to pry the top of the truck up enough to get a door open and clear away the glass. Then, with his back on the seat, the Alaskan used his big feet to push the top of the truck back in place. It was obvious the truck had been in a wreck, but you'd have to look twice to realize there was no glass windshield. Joe got in, turned the key, and the engine started!

Next they lifted the hood to examine the engine. The fan belt wouldn't turn because the fan had been pushed in, so they removed the fan belt and then the fan itself. Then they replaced the fan belt, realizing it was cold enough in Alaska to drive a short distance before the engine overheated.

This meant we would have to take a ferry directly to Seattle. The weekly ferry was leaving early the next morning, but the dock was only 16 miles away. Joe felt we could make it that far even if we had to stop along the way to let the engine cool. He drove the truck back to our park, and with Tim's help mounted the camper on the damaged truck. Then he hooked it to the trailer.

We didn't have a reservation, so we needed to be first in line at the ticket office to have the best chance of getting on. The storm that began the night before became worse again as night fell. We planned to park at the dock until morning and be first in line. Driving 16 miles without a windshield meant we were pelted with icy snow and hail all the way. Tim was yelping, "Ouch! Ouch! Ouch!" until Joe told him to cover his head with a blanket and shut up.

My job was to try to keep Joe's glasses clear enough for him to see. This required wiping first one glass and then the other so he could see enough to keep the truck on the road. There was no other traffic. By inching along, we made it to the dock and parked, ready to buy a ticket. We hurried inside the trailer, grateful for its protection. If we didn't get on this ferry, we'd have to wait at least one more week, and Joe no longer had a job or a way to get to one.

The next morning we were told that getting there early didn't help. Everybody except us had reservations. They would take us as a standby *if someone didn't show up* or there was extra room. It was a nail-biting time as we watched the ferry crew load cars and then a large number of trucks that appeared one after the other that morning. It didn't seem there would be any room for us. Disappointed and heartsick, we were about to go back to our trailer when the man in charge of loading called Joe. After taking many measurements, he said he *thought* he could load us. We barely squeezed in.

The ferry went to Sitka first, where several trucks in front of us had to be unloaded. For them to do that, we had to back off and up a big ramp in reverse gear. With no radiator working, Joe had to work fast before the engine overheated. He barely made it and quickly shut the motor off.

It looked as if everyone in Sitka was watching this strange combination backing up the ramp. Few RVs went to Sitka in those days. Again, we had to wait until they loaded on more cars and trucks going to Seattle. We needed the time to allow the engine to cool down enough for Joe to pull our rig back onto the ferry. Again, we were the last to load.

By the time we got off the ferry in Seattle, Joe had figured out a way to cut the fan blades to about half their normal length so they had room to turn without scraping anything. This allowed the pump to provide just enough coolant for cold weather driving.

We could have bought a truck in Seattle, but there would be a heavy sales tax. Joe insisted we could make it to Oregon where there was no sales tax. It meant driving slowly, but we needed to do that anyway since we had no windshield. Ever wonder what would happen to the bugs that commit suicide on your windshield if you didn't have one?

Joe instructed, "If we pass any police, smile as if we were in a heated truck. I don't want a ticket for driving without a windshield."

In Portland, we purchased a one-ton Chevrolet. As we headed south, I thought about the past four wonderful months in Alaska. Before I met Joe, I thought real Alaskans were Eskimos who live in igloos, traveling on dog sleds. If that Alaska ever existed, it no longer does. Yes, many Eskimos do live in isolated villages, but their homes are permanent frame buildings mostly of wood. The igloo is only a temporary winter house and the tent a summer house used only when they are "on the hunting trail" that is now usually reached by sleds driven by motors, not pulled by dogs.

Some traditions have mixed into the modern world. Eskimos still carve ivory into trinkets and make fur-lined boots and jackets to sell to tourists. In winter, they fish for food in holes cut into the ice, but they also eat prepared food flown into their village by small planes. They still love to dance to the rhythm of a walrus hoop, but it is done more often to get money from tourists.

Eskimos are only one type of Alaskans. The early Alaskan settlers, the Matanuska Valley colonists, and the newcomers may not have been born in Alaska, but their offspring usually have. People who have made Alaska their permanent home proudly answer to the name of "sourdough."

What Is Alaska Like?

Alaska is the hemlock which sighs in the breeze
And the lemon-lime leaves of deciduous trees.
It is snow-covered mountains under soft blue skies
And a shadow on snow where the proud eagle flies.
Alaska is the flowers—so gay and so bright—
That flourish in summer when there's almost no night.
It's the big grizzly bear and his fierce brown brother
And the little black cub with his curious mother.
Alaska is the cities where tall buildings stand
And the small log cabin on a homesteader's land.
It's the trail of a dog sled across frozen snow
To some remote village where the Eskimos go.
Alaska is king salmon, and the little songbirds,
And caribou that migrate in thundering herds.
It's the bleached white antlers that hang over each door.
Alaska is all these and a million things more.
(written by Kay Peterson, September 1972)

CHAPTER 13

RV Pioneers (1973—1977)

TRAVELERS WHO RACE against the calendar going from one tourist attraction to another miss the essence of the travel experience. They remind me of the parable of three blind men who were led to an elephant, allowed to touch it, and then asked to describe what they thought an elephant was like. One took hold of the tail, another the trunk, and the third examined the foot. Their conclusions were not the same; yet, each spoke with authority believing he now knew what an elephant was like.

I did not want our travel experience limited to what we could see from the window of our pickup. Some travelers stare at the Statue of Liberty, watch the shifting desert sands, and gaze at the Grand Canyon, but all they have are the bragging rights of a blind man who thinks he knows what an elephant is. You cannot know America until you have examined it from every side and listened to its heartbeat. That was my goal, no matter how long it took or how many detours we made.

After we returned from Alaska, Joe went to work in Washington, and Tim went to school there. His dislike of changing schools was becoming more apparent, although his grades never suffered. Sometimes he was ahead of his classmates, and other times he was behind. When that happened, Joe helped him catch up on math, and I helped on English grammar. History and geography were no challenge.

We had replenished our savings over the summer, paid cash for a better truck with the insurance money, and the nuclear plant where he worked in Washington had a lot of overtime benefits. When summer drew near, Joe suggested we spend it mostly traveling. We asked Tim where he would like to go.

He said, "Jeff is living in Atlanta. Can we visit him?"

"Sure," Joe replied. "I need to see how my sister's family is handling Walter and Lena living with them. We can swing through Atlanta on the way."

Tim was elated. When we reached the Schmidt home, we found that, like Tim, Jeff had not made another best friend. The boys were delighted to see each other; the Schmidts suggested we let Tim stay there for the summer. He would start high school in the fall.

We parked on his sister's land again while Joe worked in Florida after all. In mid-August we picked Tim up, said good-bye to Jeff's family, and headed west. Joe had checked with union agents, and there were only a couple of places he could be sure of working for the entire school year. One place that guaranteed a year's work for a tramp was another copper-mining town in the small town of Morenci, Arizona. I pictured a place like Oracle where we had so much fun. Morenci was different. Tim looked at the dismal, bare dirt landscape and asked, "Is this where we are going to live?" I detected tears in his voice.

I felt like crying, too. "It's the only place where Dad knows the job will last all school year. We'll travel wherever you want during the summers."

"Why can't we go back to California?"

"We sold the house and that's the only other place I can work." Joe explained. "There are only a couple of trailer parks in the entire area. I checked and they all have long waiting lists."

We should have kept our house and, when Tim got in high school, we could have returned each fall, traveling only during the summer. That was what we originally planned. It didn't matter that *some* children enjoy traveling and don't mind changing schools; the reality was that any allure of travel had worn off for Tim.

The next day, I drove Tim to school to sign up for his classes, which would start the following Monday. I hoped he would find it easier to make friends in this small school. I tried to forget how long it usually took in a new place and remembered the loneliness of my own school days.

When Joe came home from his first day on the job, he wasn't happy either. "It's hot as hell on the jobsite," was all he said.

I'm not sure which of the three of us felt the most miserable, but I think it was me because I was the instigator of traveling with children. I had been blinded by what seemed a perfect childhood to me. Now our RV truly was our only home and returning to our former life was impossible.

On Saturday, I took Tim with me to set up a post office box and see if there was anything of interest in town. Two letters were waiting at general delivery. One was to Tim from Jeff and the other to us from his parents.

Our letter suggested flying Tim to Atlanta and letting him go to school there. It seems that, after we left, Jeff told his parents how much Tim hated trailer life. At the time, he didn't even know what awaited him in Morenci!

Today, more full-timers travel with school-age children because home schooling is more acceptable. Computer and correspondence courses have advantages not only for travelers but also for children who have trouble fitting in. The "tramp children" I've come to know are bright, mature beyond their years, and not involved in drugs because there is no pressure from school friends. Had home schooling been as available then, I think Tim might have been happy traveling, and tramping would have been much easier.

That evening Joe called Andy and they came up with a plan. Tim would stay with them during the school year and Jeff would travel with us next summer. We would renew our contract on an annual basis, and either party could cancel the deal if it wasn't working out. I wish Tim had told us or Jeff's parents how he felt when we were at their house. Making arrangements would have been a lot easier there than long distance.

We knew the Schmidts shared our philosophy about raising children, how much freedom to allow them, and their need to earn their own spending money. That was more difficult in our lifestyle, but since Andy managed a restaurant in Atlanta, he found jobs for the boys on weekends. In the years that followed, they advanced from cleaning to dish washers, bus boys, waiters, and then cooks. Jeff loved the bakery and made that his career.

As it turned out, spending those four years with the Schmidts gave Tim both the stability he craved and his first career. He continued up the ladder to manager, teaching cooks and managers, and then owned a restaurant in partnership with a buddy. It was successful, but Tim was tired of the demands of the restaurant business. He wanted to try something else.

Next Tim and his wife sold antiques on e-Bay, plus having a small antique store. (On a visit there, I learned how valuable my grandparents' things were that I had given away or sold to a secondhand dealer.) Then Tim started working as a used car salesman and ended up as a top executive with a successful multi-state dealership.

There is no telling what he will try when he gets tired of that. I believe the experience of traveling opened his eyes to the world and its many opportunities. He has been successful at everything he did, and his three children are following in his footsteps.

During the summers, Joe worked only a week or two at a time so we could show Tim and Jeff as much country as possible, especially historical sites.

One summer experience was a valuable lesson for both teen-agers. We stopped in Washington to visit Donny, Joe's Alaska buddy. Donny had a son the same age as Jeff and Tim. Donny found a job for all three boys bundling, carrying, and throwing bales of hay on a truck. Unlike their jobs in the restaurant, this was hard physical labor outside with the hot summer sun beating down. They worked most of the morning and then told the farmer they were going home.

Donny's son couldn't understand it. "You don't want to leave now," he said. "It'll be lunch time soon and there will be tons of food and pies and cake."

His argument made no impression. One morning of hard physical labor made them both realize that was not how they wanted to make a living. It inspired them to have higher expectations.

When we made the decision to let Tim stay with the Schmidts during the school year, we did not realize how much easier tramping would be. The first thing Joe did was quit the Morenci job. We hated being there as much as Tim did.

Making it easier to travel with children is only one of the changes since our RV pioneering in the 1970s. Handling your own money was another. It was almost impossible to cash out-of-state checks even with identification. Bank tellers and store managers eyed travelers with suspicion. Credit cards existed, but people were cautious about using them.

In Florida, Charley was refused by two banks when he tried to cash a check. When the third bank refused him, he lost his cool. "What do you people call sufficient identification?"

The woman explained, "Your driver's license doesn't have your picture on it. How do I know this is you?"

Charley dug into his billfold and pulled out a copy of his service record. It said he had dentures and gave the serial number. While the teller was trying to figure out how this proved anything, Charley jerked out his upper plate and laid his teeth on the counter in front of a suddenly sick-looking teller.

"The same serial number is stamped on it. That should be proof enough."

One of the things full-time RVers had to do in those days was to think outside the traditional box the way Charley did.

There were no cell phones; RVers were forced to find and use pay phones. We stopped at a store that advertised a special on engine oil. Joe bought a case plus some other things he needed. When he got to the counter, he realized he didn't have enough cash, so he asked if the cashier would take a check. As she was preparing to cash the check, she noticed the address was from out-of-state and insisted on getting a telephone number.

Rather than try to explain to someone who thought everyone had telephones, he made up a number and gave it to her. She wrote it on the check and he carried his purchase to the trailer. While putting the oil in an outside storage area, he began thinking what a bargain it was and decided to go back and get another case.

There were two cashiers, but one was tied up with a big order so he went to the second one. While she was ringing up his order, he realized it was the same cashier he had just used. He couldn't remember what phone number he gave her the first time but thought it wouldn't matter because cashiers handle so many checks.

"Phone number, please?"

He was sure he used our old area code, so he rattled those numbers off along with seven more numbers. She paused, looked up at him, then opened her drawer and took out his other check. "That's not the number you gave me last time," she said suspiciously.

"Oh, yeah. That one was my business number. This is my home number." He used his old trick of making fun of himself. "I'm such an idiot. Why do I do such stupid things?"

"Don't worry," she sympathized. "We all do dumb things at times."

In those days, all RVers had their own check-cashing stories. Mine happened in a grocery store in the town next to the bank we had used during our travels. After spending the better part of an hour selecting a cart full of groceries, I stood in line 15 minutes waiting my turn to check out. I told the clerk I was paying for them by check.

"Do you have identification?"

"Yes," I said with confidence.

After she finished ringing them up, I wrote a check while she bagged the last of them.

She looked at the check. "You didn't say it was an out-of-state check," she accused. "We only accept local checks."

"It is local," I said pointing to the bank address printed on the bottom of the check.

"It says up here you live in New Mexico." Her finger was underlining my printed address at the top of the check.

"That's a mail-forwarding address," I assured her. "I live in an RV and we travel, so I don't have a permanent address." As soon as the words left my mouth, I knew it was a mistake.

Glaring at me, she called the manager. Neither of us spoke, but I heard impatient shuffling of feet and whispered comments from the line that had formed behind me.

A young man appeared, wearing a fake smile and a big manager's badge.

"She has an out-of-state check and didn't tell me," the clerk said.

"It isn't out-of-state," I pointed at the bank address.

"Her address is New Mexico," she interrupted.

While the manager examined the check, I explained again that it was where we got our mail, not where we lived.

He seemed more confused. "Where do you live?"

I explained as patiently as I could with the comments behind me turning from whispers to rude remarks. "My home is a travel trailer, and we live in it."

"When you aren't traveling, where do you live?"

I decided to play his way. "At the trailer park down the street."

"Oh. Why didn't you say you live here now? Let me see your driver's license."

I knew I'd lost this battle. It was easier to give him my California license than try to explain.

"What are you trying to pull? All your addresses are different: a bank in Arizona with a New Mexico address and a California driver's license. Then you try to tell me you live here. Get out and don't come back. We don't need your business." He handed me my license and the check.

As I walked away trying to ignore the laughter and remarks made about me, I told myself that somebody had to put away all those groceries. I hoped it was the clerk.

What people didn't understand was that communication was difficult for RV pioneers. We used the address of whatever family member or friend was handling our mail. We could only write directly to people who stayed in one location. When I wrote to travelers, my letter went first to the person who handled their mail.

We all had someone collect, throw out trash mail, and forward the rest. RVers used different methods for forwarding. Some had their handler wait until there was enough to send all in one large envelope; others crossed off the address and wrote the forwarding address on each letter. This saved forwarding postage but did not guarantee that all the letters would arrive before you moved on. Lost mail was common.

The people who call today's post office system "snail mail" have no idea what *real snail mail* was like. It could be held up for weeks if your forwarder was busy or waiting for enough mail to accumulate to make it worth a trip to the post office.

We paid cash whenever we could. When insurance payments were due, we sent a check without waiting for the billing statement to arrive. Again, this is ancient times—1970s—before cell phones, computers, e-mail, iPods, and whatever other technology someone is inventing as I write this.

Using a pay phone was a nuisance. If you didn't have enough correct coins, the operator would shut down the phone—sometimes in the middle of an important conversation. Most RVers made calls once or twice a week. The person we contacted was supposed to relay our message to others, but some people were more dependable than others. Many of us prescheduled phone calls.

There were pay phones at every shopping center, gas station, and restaurant in those days. Many were enclosed in phone booths, but others were attached to the outside of the building. Even with a phone booth, you often stood outside in blazing summer sun, rain, wind, or freezing cold waiting for the person ahead of you to get off the phone.

Another problem in the 1970s was the lack of repair facilities for RVs. Whenever possible, we went to the dealer we purchased from or the factory where it was built. Joe could fix almost anything wrong with the truck and had repaired broken down washing machines, dryers, vacuum cleaners, and other equipment.

One Saturday morning when we found our refrigerator wasn't working, Joe decided he could fix it. I handed him the repair manual. "This looks complicated," I said. "Maybe you should read all the instructions before you decide if you can fix it."

He opened the manual. "Why do they give us a repair manual if we aren't supposed to use it? Don't worry. I can handle it." He sounded confident.

Step #1: *"To clean the flue, you must gain access to the back of the refrigerator."* That was followed by diagrams that meant nothing to me. Joe skipped over them and was opening his tool box.

"Did you understand *all* the instructions?" I asked. I knew he had not read beyond the first step.

"The manual explains it step by step," Joe replied with an air of authority.

Step #2: *"Remove all the screws that tie the credenza to the wall of the refrigerator."*

Credenza is a fancy name for a storage closet above the refrigerator. Ours was stuffed with things we might need if we remembered where we put them. To get to the screws, everything had to be removed. I helped him empty the compartment. It was a good review of what was stored there.

Step #3: *"Remove the partition between the credenza and the refrigerator."*

Since this partition was actually the credenza floor and was also secured to the outer wall of the trailer, more screws and rivets had to be taken out. He put these in a separate pile so he knew where they went when it was time to reassemble.

"Joe, this is getting more complicated. Don't you think you should read through *all* the directions to see if you can do this?" I suggested again.

He suggested this might be a good time for me to go for a walk, but I would never desert him in the middle of a complicated project.

Step #4: *"Remove the screws and rivets that tie the side of refrigerator to the pantry wall."* In order to get to those, we had to remove everything on the pantry shelves.

"Now, aren't you glad I stayed?" I said somewhat smugly.

He did not answer.

"Hey, look at this! It's that package of cornbread mix we lost last year," I announced. Joe was unimpressed.

I will not go into all the details, except to say things became more complicated. Finally, he was able to pull the refrigerator into the middle of the aisle where he could reach the back of it. He found the flue and repaired the problem. Replacing the refrigerator was a reinforcement of why he really should have taken my advice about reading the instructions before he started.

Joe is a person who learns from his mistakes. Months later, we were in Indiana heading to another job when the door lock broke, causing the door to fly back and hit the side of the trailer. This time, Joe investigated it before attempting repairs. He discovered when he closed the door there was a gaping crack along the top and the bottom of the door resulting from its encounter with the side of the trailer. The latch was broken, too.

He went searching for a phone. Remember, this is still prehistoric times—probably 1976 now. He found a service-repair center in a town on the way to his next job. When he explained the urgency to the service manager, he was told they would fix it immediately if we were there by 8:00

a.m. Friday morning.

We were parked outside when the shop opened Friday, but the manager explained there was a slight delay. They had an emergency job for a *regular customer* to do first.

At 10:00 a.m., Joe went back to the manager's office. They were almost ready for us.

At 11:00, he made another trip to remind the manager we had an 8:00 a.m. appointment.

At noon, they closed for lunch.

At 1:15, they were ready for us.

The mechanic assigned to us was young but friendly. He assured us we didn't need a new door (which he did not have). He assured us that adjusting ours was no problem. After he finished adjusting it, the gaping crack at top and bottom was evenly spread around the entire door. He assured us it would work fine as soon as he replaced the lock.

It seems they didn't have the right lock in stock, but he assured us he could repair our broken one. I liked this young man's confidence. He reminded me of Joe.

At 4:30, the manager left for the weekend. Joe went back to the shop area to see how our mechanic was doing. He was still repairing the lock.

By 5:30, our mechanic was the only one still working, but he announced the lock was now repaired. Joe pointed out that the door wouldn't close.

"Trailer doors are hard to shut," the mechanic assured us. "You have to slam them." He grabbed the handle in preparing to give the door a push. The handle broke off in his hand. It is sad to watch a mechanic cry.

He assured us he could put on a new handle. Deciding it was too painful to watch, Joe suggested we go for a walk. At 6:15, we returned and found the mechanic kneeling beside the trailer. I thought he was praying, but he was filing down the latch plate.

"The lock doesn't fit right," his voice was cracking a little.

"It probably just needs a little adjustment," Joe said kindly. "Why don't you go home? I can fix it later."

With a sigh that was more like a sob, the mechanic forced the door shut by shoving hard and then leaning all his weight against it.

We settled our bill and drove away. It had been a long day sitting in waiting rooms, reading old magazines, so we stopped at the first rest area we came to.

Opening the door of the trailer proved to be easy enough, but we couldn't close it from the inside. After some experimenting, we found the only way to close it was for one of us to stand outside and lean against it. Neither of us was willing to spend the night doing that.

Joe insisted there must be a way to close it from the inside. He grabbed the handle and gave the door a mighty tug. The handle broke off again. It is *really sad* to watch your husband cry.

Finally, Joe secured the door with stretch cords, duct tape, and bailing wire. There was no way to lock it; we were too tired to care.

Saturday morning, I was in much better spirits because Joe had a new plan. Since we were happy full-timers no longer worried about school deadlines, we could take a detour to the factory. We could easily make it there by Monday, have the whole door and lock replaced, and only be a day—or maybe two—late arriving at the new job.

When we arrived at the Airstream factory late Sunday afternoon, we learned they had a free trailer park complete with hookups for customers to use. Wonderful. Other late arrivals kept pulling in well into the night, filling all the spaces around us. In talking with neighbors, we learned everyone—except us—had an appointment for some time the following week.

It didn't worry us. We assured each other that emergency work, like fixing a door that wouldn't close, would receive priority. We were wrong. The next morning we learned our problem was not an emergency, and appointments were made three months in advance. The service manager was the friendly type who said we could stay right there, even if we had to wait three months.

"We're on our way to a job. Can't you squeeze us in? Our problem won't take long to fix."

He promised to see what he could do.

I felt sure he would squeeze us in before the day was over. He didn't get to us that day. Or the next. On the third morning, a small circus pulled into the trailer park. The circus owner lived in a trailer that needed repair work, and he didn't have an appointment either.

Along with his own trailer, he brought his performers in a bus, his maintenance crew, a trailer for employee sleeping quarters, a number of dogs, three tigers, two lions, and an elephant. Whatever his problem was, it qualified as an emergency. (I think it helped when the elephant knocked over a lamppost.) They didn't even have to stay overnight. However, some people were upset because their appointments weren't being met on schedule.

Our job commitments were never made in concrete because of these unexpected interruptions. If you missed out on one job, there was always another one. The '70s and '80s were peak years for construction tramps. I stopped making deadlines. We wanted to take advantage of opportunities that came our way, so delays became more fun than frustrating, even though we had to work to pay for our travels and other commitments.

Our friends, Jim and Marge, were retired full-time RVers totally free to change their plans on a whim. I hoped to be like them someday. RVers who live on a limited income often "boondocked" or "dry-camped" as some called it. This means parking, usually overnight, anyplace that looks quiet and convenient. Many RVers still do it, but it was more common during RV pioneer years.

One night they pulled off onto a deserted dirt road. When they awoke in the morning, they heard angry voices discussing the presence of their trailer. Looking out the window, Jim discovered that the dirt road he picked led to a lettuce field that was now swarming with migrant workers. The bus that brought them was deliberately parked to block their trailer from leaving.

Jim went outside to talk to an angry foreman. Pretending not to notice his hostility, Jim said cheerfully, "My wife's making coffee. If you can leave your men for a few minutes, will you join us? I suspect she's making cinnamon rolls, too."

The foreman glared, pretending not to understand English. Jim continued in the same cheerful voice. "I can bring a cup of coffee to you, but my wife thought you might like to see how we live. That's the only home we have. Guess you might say we're migrants, too."

Whether it was this comparison or the usual curiosity about trailers, the foreman followed him inside. As Jim pointed out the conveniences of a trailer, the hostility disappeared and he began asking questions (in English) while they drank coffee and ate cinnamon rolls. The foreman announced he had to get back to work. He was expecting Jim to ask him to move the bus. Jim had a different idea.

"I've never picked crops, but since I have to stay here until you finish work, I'd like to try it. How about a job?"

The foreman stared. Then a grin spread over his face. "We pay by the box. You won't make much."

"That's okay. The way I figure it, a man should try everything at least once."

"What about your wife? What'll she do all day?"

Jim put on his favorite baseball cap. "She's not much for gardening. This is our home. She'll do the same things as always."

When lunch time came, Marge set up a card table under the awning and served ice tea to everyone. Jim gave more curious people a tour of the trailer. Late that afternoon, but while the laborers were still working, Jim had had enough and told the foreman he was quitting. He took the few dollars he had earned and a big head of lettuce as a gift.

"It'll be dark soon," the foreman said. "You might's well stay the night."

When the migrant bus arrived the next day and found their trailer still there, they parked the bus well out of Jim's way.

You never know when you wake up in the morning what adventure will happen that day. For instance, we were on Interstate 10 heading west to a job in California. Lynn and his family from Alaska were heading east to a job in Phoenix. Lynn saw us first and blinked his lights.

Joe said, "Hey! That was Lynn!" He pulled off the highway.

Lynn had pulled off on the opposite side. All of us ran to embrace each other.

"We're heading for a county park about 10 miles ahead on this road." Lynn said. "Are you in a hurry, or do you have time to visit?"

"We're never in too much of a hurry to talk," Joe replied. "I'll have to find a place to cross over, but we'll meet you at the park as soon as we can."

Three fun-filled *days* later, we each pulled out of the park headed in different directions again. Unexpected encounters are great.

I have many wonderful memories of our years of full-time RVing and working. While these adventures were taking place, my writing career blossomed. It began when I submitted an article to *Trailer Life* magazine about traveling with a child, (the same article Lynn had read in Alaska). Because there was so little information available about full-time RVing, I decided I knew as much as most RVers. I was already taking a correspondence writing course as a hobby. When *Trailer Life* published my story without changing anything but punctuation, it inspired me. I submitted a query about writing a series of articles on how to travel full-time.

They agreed to accept it if I gave them "all rights," which meant I could not use the same information anywhere else. I thought that after I had enough material I might write a book on full-time RVing. We had met enough curious people to believe there was a growing interest.

So I rejected their offer and submitted the same query to another magazine, *Woodall's Trailer Travel,* who immediately accepted my offer and wanted a column every month. The editor agreed to purchase only "First North American Rights," which meant I could republish any articles at a later time. Much of those basic full-time RVer columns ended up in one of my eight books.

Listing me on the magazine masthead established me as a professional writer. My columns must have been popular because each magazine had complimentary letters about them in the "Letters to Editor." It was also the beginning of a friendship with the editor and my copy editor. For the next eight years we made an annual visit to Chicago. The editorial staff sat in our RV parked on a Chicago street while we told them about our travels.

Along with the RVing series, the editor wanted me to write a series called "Kay Peterson's America." I picked the places to write about. This opened doors to experiences we would never have had without my writing credential to pave the way. (You will read the most memorable experiences in other chapters.)

At the same time, the magazine was getting many technical questions, so the editor asked me to switch from the how-to column to a question-and-answer column. I readily agreed. I had already covered the RV lifestyle, and this new column was one Joe and I could do together. He answered the technical questions, and I answered the lifestyle ones.

Both of us were listed as contributing writers when the magazine was sold to another company that soon stopped publication. I think their goal was to eliminate a competitor. The editorial staff went to work for other magazines, and I lost contact with them.

My years of writing for an RV magazine came at a time when people were thinking seriously about selling their homes when they retired and living out the rest of their active lives on the road. Yet, many wondered if it was possible to do. My monthly articles answered the fears of those people.

My book, *Home Is Where You Park It*, was the first book ever published on *living full-time* in an RV. The first printing in 1977 was published by Follet Publishing Company in Chicago. We thought they would market it, but they had other books coming out at the same time that they thought the public was more likely to buy. Marketing budgets are always limited.

The publisher sold very few books. We bought my own book from them at a discount and sold them through my magazine columns. Joe also sold them to construction tramps on jobs and to interested people we met as we traveled. When the first run sold out, Follet decided not to republish. They returned my copyright and we continued publishing it for 15 more years with a different cover and periodic updates.

When the book first came out in June 1977, we picked up 200 copies on our way north after attending Tim and Jeff's high school graduation. This would be their last summer of RV travel.

We had signed up to attend a national Airstream rally, and I was hoping to sell a few books there. I had written the seminar director, told her I was the full-timing columnist for *Woodall's,* and asked if I could give a talk on full-time RVing.

It was mind-boggling how a few thousand Airstreams were set up in perfect alignment across a college campus. They had rented small buses to take people from their rigs to the various activities. Joe volunteered to drive one of the buses and had a great time joking with his passengers. Mostly he had the same passengers, but he always reminded them about my talk on full-time RVing and gave daily accounts of the chess tournament he had entered.

The other chess contestants were retired military officers and business executives, and they all seemed to know each other. To their astonishment, Joe was beating them, one by one, and having a great time doing it. He confided to me that so far he had not encountered anyone he considered an excellent player in spite of their boasts to each other.

At the final round, after all the eliminations, Joe was playing a high-ranking military officer who obviously expected to win. The room was full of his friends who had competed and been eliminated. They were the cheering section for the officer. When Joe won the tournament, the military man said, "What did you say your former career was, Peterson?"

"I'm still working. I'm a union construction electrician who travels," Joe beamed proudly.

The room became deadly quiet. The officer he had beaten walked away, ignoring Joe's extended hand. "That room sure emptied fast! Only one man shook my hand and congratulated me. I guess they couldn't accept that a common laborer was better at chess than they were."

The next day on his bus route, he brought along the trophy and showed it to the passengers at every stop, no matter how many times they used his bus that morning. Every time they clapped—probably at his enthusiasm. Joe kept that trophy prominently displayed for the rest of his life, even after it became tarnished beyond repair. I think it meant even more to him because of his opponents' attitude.

On the last day, my talk was scheduled for 2 p.m. I had attended a number of seminars about places to travel and new equipment that was available to make travels easier. There was nothing about living in an RV. It was as if everyone only used their trailer to attend local rallies and this annual event.

The seminars were held in an auditorium with theater-type chairs and a professional sound system with microphones for the speaker and the host.

I assumed I would speak there also, but when I finally cornered the seminar director, she said, "Didn't anyone tell you? Mr. X is a professional speaker, and he'll be using the auditorium at the same time as your little talk."

"Then where will I give my talk?" I wanted to use the demeaning word "little" in a sarcastic way, but I was afraid she might be right. I might not have any audience.

"The staff put chairs in the gymnasium. It's in another building. Look on your program map."

When I told Joe, he said, "That's in a different location. Maybe we better take a look at it."

We arrived while everyone was at lunch. We were dismayed that there was no sound system. In such a large open room with high ceilings, Joe was afraid my voice either wouldn't be heard or would have an echo. My attention was on a row of chairs in front of a table with a small lectern on it. "I guess they don't expect many people." There were 12 chairs.

Joe was angry. "How do they know how many are coming? I know some of my regular bus passengers will be here. Let's check the other rooms and see if we can find more chairs."

We rounded up another two dozen chairs and placed them behind the first 12. Then Joe rearranged them with eight in each row so we could have more rows. Upset and nervous, we returned to our trailer to wait.

As we started back to the gymnasium, I said, "I was going to bring a box of books, but I guess I'll just take one to hold up and show people."

"If you want a box, we'll take a box," Joe said. I could tell he was angry. "I'll carry it over there, and I'll carry it back if I have to. I bet everyone who comes will want a book. Everybody I talk with about what we're doing is really interested in knowing more."

He was always there for me, beside me, ready to fight for me. He was also my biggest salesman. We held hands as we walked across to the building. People smiled when they saw us; that was encouraging.

When we arrived, all the chairs we had put out were filled, and people were bringing in chairs from other rooms. By two o'clock, the room was filled with people. Many were sitting on the floor and others were standing against the walls. The gymnasium windows on the groundside were open. People were sitting *outside* with their heads stuck inside to hear as much as possible.

The seminar director came in flustered. "My goodness! I never thought so many people were interested in *living* in a trailer. I think we should reschedule this for later."

"No!" yelled the people who heard her. "We're here. We want to hear this."

"These people should be in the auditorium," she whispered to me. "There's almost no one there."

I pretended not to hear.

Joe took command. "My wife isn't very tall. Some of you can't see her. Kay, why don't you go up on that balcony and speak from there?"

I was in a daze but followed his advice. By the time I reached the balcony, the seminar director had disappeared. The crowd looked up, patiently waiting for me to speak. From above them, my voice would be more easily heard, but it was annoying dodging the basketball net that kept bumping my head. I had no notes and was suddenly nervous.

As soon as I started to talk, all nervousness disappeared. I began by giving them my credentials since I had no host to do that. Then I wove in stories to illustrate basic advice. The book was fresh in my mind, so it was easy. When I finished, the applause and standing ovation brought tears to my eyes.

Joe came to my rescue again. He went to the lectern and began talking about my book, adding that we had a few copies for sale, and his wife would be happy to sign them. While I was trying to sign books, people crowded around me, thanking me and asking questions. After Joe was sold out of books, he began answering questions, too. I don't know if the seminar director ever came back to check, but people kept us there talking a long time.

Finally I said, "We need to put the chairs back." Somebody laughed. Then I realized all the chairs and lectern were gone. The only thing left was the table I had been using to autograph books.

To me, speaking is more powerful than writing because you see the reaction to your words on people's faces. A sad story brought tears; making fun of myself or Joe brought laughter; making points and seeing heads nod in agreement was gratifying. Even without all the props professional speakers have, that "little talk" was the beginning of a speaking career that lasted for more than 25 years at RV shows, educational seminars, and club rallies across the country.

Our goal was to pave the road for easier access to people who wanted to enjoy the RV lifestyle. Yet, it was more than that to me. I wanted to encourage and inspire people to follow their own dream.

Joe and Kay Peterson gave full-time RV seminars from 1977 to 2002.

CHAPTER 14

Starting the Escapees Club (1978—present)

IN 1978, LYNN Rogers, who was trying to get into the electrical trade, was able to get an out-of-class clearance to the nuclear power plant in Washington. He wrote us saying, "Where are you? You told me if I ever got on a job, you'd come and teach me the trade. Here I am."

Joe was amazed at how Lynn had talked his way step-by-step toward his goal. We headed for Washington, where a nuclear power plant was in constant need of electricians, millwrights, and laborers. Joe talked his way into being Lynn's "tool buddy," and we parked next to each other in a trailer park.

One evening when visiting, Lynn suggested starting a club for full-time RVers. I was aware of the interest in RV travel even before giving the talk at the Airstream rally the summer before. I believed RVers needed to communicate and share the solutions they discovered to common problems. Lynn suggested I write a monthly newsletter.

At the time, I was writing monthly columns for *Woodall's* magazine and for a new organization's publication called *Snowbird Newsletter*. I wasn't sure if I wanted to take on the challenge of writing and publishing my own newsletter as well. However, with both Lynn's and Joe's encouragement, I agreed to do it if enough people were interested.

As it happened, one of the questions in our file asked if there was a club specifically for full-time RVers. The two major clubs, Good Sam and Family Motor Coach Association (FMCA), had a chapter for full-timers, but they weren't very active.

In our June 1978 question/answer column, we published the question, and I answered by saying I would organize a club for full-timers if there was enough interest. As soon as that magazine was published, we had two dozen responses, some from people who didn't even own an RV but wanted to learn about it. I answered every letter of inquiry or response to our column. We welcomed those living in their RV and also those we called "wannabe full-timers."

Joe's work provided the money for a better typewriter and a mimeograph machine. My original response letter requested annual dues of three dollars to cover postage costs and asked early responders for suggestions to name the club. We used the term "club" defined as "a group of people organized for a common purpose."

Most suggestions evolved around being free and traveling without deadlines. The name we selected was *Escapees*. It seemed a perfect name to me because both Joe and I had escaped almost certain death in our earlier lives. We both had beaten the odds more times than we should have, and that is a form of escape. Now the term meant escaping from the routine rat-race of life.

The name, Escapees, was submitted by Harry Lewis along with a shortened form of SKP (say it fast and it sounds like Escapees). Harry also submitted a logo showing a window with prison bars and two hands spreading them apart. We turned his logo down for fear people would misinterpret our purpose.

Even with a different logo, the name caused a few problems over the years. We used our personal logo previously created for stationery by a wannabe artist using Joe's idea of a small house in a red wagon. It was a distinctive logo, but as Escapees grew, real artists wanted to perfect the drawing that did not have correct proportions. Except for a few subtle changes, we registered it, imperfections and all.

I never intended the Escapees Club to be anything more than a newsletter. This was at a time when the word "computer" was a mysterious machine used by big businesses. My writing was done on an unforgiving old typewriter. Now I was typing on a stencil, even more unforgiving of mistakes. We needed it so Joe could make multiple copies on a mimeograph machine we bought at a pawn shop. Like our logo, it was far from perfect, but people didn't seem to care. Communication with others in the RV lifestyle and a chance to exchange information was what mattered. One person said, "Send the information on toilet paper if you have to."

Once the pages were printed, we stacked them in piles on our floor. Then we assembled the pages, stapled them together, folded them in half, wrote an address and a short personal message on the back of each, pasted on postage, and took them in paper bags to the nearest post office. The personal message was to atone for the crudeness.

We mailed that first five-page newsletter in August to the 82 families who had joined by then. Since members began joining in July, we declared July 4, Independence Day 1978, as the founding date of the Escapees Club.

The odds were against us being more than a flash in the pan. Several others tried to start newsletters after ours. Only trade magazines or chapter newsletters backed by big corporations were successful. Certainly, the odds were against us with no financial backing except Joe's work.

Escapees grew slowly by word of mouth and notices inserted at the end of our columns in *Woodall's Trailer Travel* magazine and *Snowbird Newsletter*. I was still not convinced we even needed another publication, but from day one those who joined encouraged us to continue.

Many times in the following years, I thought of myself as standing on top of a snow-covered mountain while I molded a handful of snow into a ball called Escapees. I allowed it to roll down the mountain under its own momentum. I was amazed how quickly it gathered people, new ideas, and more goals until today we have a huge snowball and a multi-faceted organization.

People called us great visionaries. The truth is, in those early days there was no plan with three-, five-, and ten-year goals. Our snowball grew from the requests of members. After full-timing for eight years, we knew which suggestions were valid for the majority of RVers.

The first request was to meet each other in person. Less than a year after the first bimonthly publication, we held our first rally in Bakersfield, California. It was February 1979, and someone called it an Escapade. The name stuck for our future major events.

Twenty-four families, including nine children, attended that first holiday weekend Escapade. Bonding began, a philosophy of caring and sharing was established, and the hugging tradition emerged spontaneously. These were the foundation on which a simple newsletter became a support system for anyone traveling in a home on wheels.

For the first few years, Joe and I published the *Escapees* bimonthly newsletter from wherever we were when it was due. We had to support ourselves and a club with too small a membership to even pay all its printing and postage expenses, in spite of the fact that by now the dues was $10 a year. At that time, I was writing articles, editing (with my limited skill) member submissions, and publishing what had become overnight a 20-page newsletter.

At the end of 1980, *Woodall's* magazine was sold and the *Snowbird Newsletter* went out of business, erasing both of my monthly writing obligations. We now had 375 member-families—more than I ever dreamed would be interested in getting a newsletter about full-time RVing.

In January 1981, I began writing a column in every *Escapees* newsletter. It started as an editorial but in a few years was changed to "Thoughts for the Road." A new column has appeared in every issue to this day. I will continue to write that column as long as I am able.

By the publication's third anniversary in 1981, it had grown to 32 pages that we were printing on a new, modern mimeograph machine. Stacks of pages on our floor waited to be correlated, making walking difficult even though we now had a 32-foot Avon trailer. We sometimes had one or two members help us attach printed address labels and postage.

In the fall of 1982, a member took us to Radio Shack to see the latest invention called a "personal computer." We came home with a computer and printer that made my work a lot easier with fewer typo mistakes. Spell-check didn't exist, but mistakes could be corrected without retyping an entire stencil.

One more year and another member introduced us to a process called web press that allowed us to change to a seven by nine-inch booklet. We no longer had to correlate pages and staple them together. The first booklet had a photo of Joe and me house-sitting for a member.

Then we hired Susie Dunaway (Gearing) as a real editor, and each booklet for the next ten-and-a-half years had a different cartoon cover designed by Anne Harris, who also illustrated some of the inside pages with humorous cartoon RVers.

In 1994 when the booklet became a more traditional magazine, many members wrote to say they missed Anne's cartoons. I missed them, too, but by now we had a graphic artist as well as an editor. We began taking a few advertisements to help pay for more computers and staff. Before the end of that year, we were publishing a 72-page *magazine*.

While continually improving our publication, we were also implementing new ideas members suggested. Immediately after that first Escapade, members wanted more than an annual get-together. We planned a second rally in the summer of 1979 in Michigan and a third one for that fall at Lead Hill, Arkansas.

The one in Michigan became memorable. A member invited us to use their property, although they couldn't attend. At their suggestion, we announced if anyone had trouble finding the land to check at the local police station.

Curt and Noreen Osborn were vacationing in the area and offered to scout it out ahead. They found it completely overgrown with brush and trees and immediately contacted us. Joe was on an overtime job in Washington and we planned to work until the last minute. After Curt called, we left the next morning for Michigan. In the meantime, we asked Curt to see if he could find some place nearby to hold the rally.

When we arrived at the park where the Osborns were staying, they had found a nearby fairground we could rent. There was no way to reach people who were on the way, so we did the next best thing. Joe and I drove to the land and put up signs telling people the rally was moved to the fairgrounds.

In case our signs blew away or were torn down by jokesters, we went to the police station, explained the situation and asked if we could put up a sign because members might inquire there. They were very cooperative and even provided a large sheet of paper for the sign and a marking pen. Their attitude changed drastically when Joe was hanging my homemade sign:

Escapees
Change of Plans
Meet at Local Fairgrounds

"Hey! I thought you said this was a camping group?" His voice was cold as steel with no sign of his friendly smile.

We showed him our newsletter and logo and tried to explain. He grabbed the newsletter. "Wait here. I'm going to make a phone call."

His assistant entertained (or was it detained?) us until he returned. He was smiling again. "What made you pick *Escapees* as the name of a camping club?"

Can you imagine how difficult it would have been to explain if we had used Harry's logo of two hands separating prison bars?

The next request was for home-base parks where members could stay between trips or permanently if they became ill and could no longer travel. From 1980 to 1983, Joe and I, with the help of dedicated members, found land and supervised the building of two Escapees Co-Op home-base campgrounds. The first was in Casa Grande, Arizona, and the second in Lakewood, New Mexico. This meant Joe was unable to work at his trade during this time.

Although we raised the dues and had more members, we were eating into our savings from the previous eight years we tramped. It worried Joe to see our money decreasing each month. I reassured him we would stop whenever he wanted to return to the tramping lifestyle we both loved. We had never intended to get this involved, but members kept demanding more parks, and now they wanted their own mail service. Their demands were realistic, but it was a struggle between our own desires and living up to the responsibility we assumed when we started the club.

In spite of our dwindling savings, we both enjoyed working on the parks with a group of volunteers who had different skills. We have many memories of that three-year period, but the one that stands out from all others is what happened when we were building the first Escapees Co-Op in Casa Grande, Arizona.

Joe was general supervisor as well as head electrician. We made mistakes because we were not qualified to build an RV park with nothing except volunteers and learning by trial and error. When the power company first came by and offered to put our electric in for a fraction of the normal cost, Joe refused. They came back a week later with what appeared to be a real blessing. They would pay for and install all 120 meters *free*.

Joe realized how much money that would save with building costs and felt we had achieved the good will of the county. Later, he said he should have examined what seemed like a very generous gift more carefully. Now the power company, instead of the members, owned each meter and could charge whatever they wanted to turn on or off the meters when a lot owner was traveling, and each person paid a minimum monthly bill. Joe apologized for the mistake and warned all future co-op builders to buy their own meters and pay only one electric bill. This meant managers had to read meters as people came and went, but it saved the co-op members a lot of money.

There was only one serious accident. It occurred when volunteers were putting in the ground rods for each meter. The clay ground was very hard, so it was a chore driving ground rods with our homemade ground-rod driver that consisted of a sledge hammer with a ¾-inch pipe welded to it for a handle. The guys took turns placing the pipe handle over the ground rod and slamming it down as hard as possible. Each downward thrust drove the ground rod a fraction of an inch further into the ground.

Joe was the only one who had driven ground rods a thousand times before, and he was the only one who got hurt. He raised the primitive driver too high so it became disconnected from the ground rod. His downward thrust caused the ground rod to come between the pipe and his bare hands, tearing up both hands.

I drove him to the emergency room in town, where, luckily, a plastic surgeon that specialized in hands was in the hospital and on call. He and another doctor did extensive repair surgery on Joe's hands under local anesthesia. They felt that, when completely healed, Joe would be able to use his hands as if they had never been torn. (They were right.)

The next day, Joe was scheduled for a dressing change, so I got behind the wheel of our pickup. He made a statement that will live in infamy. "I'll drive. I can drive better with my elbows than you can with both hands."

Joe liked to be the leader of the highway pack. Today, he was behind a slow dump truck (by slow, I mean going the speed limit on the back road to town). There were no oncoming vehicles, so he passed the truck, though we knew from previous trips this was where the police liked to "hide" (wait for someone to break the law). But when the driver in front of you is going too slowly for comfort, what is an impatient driver supposed to do? Joe passed the dump truck, which required exceeding the speed limit *and* crossing a double yellow line almost in front of where the police wait. As soon as he finished the passing maneuver, I heard a siren and saw flashing lights behind us.

Joe pulled over; I got his license out of his billfold and, holding it between the only two fingers that were not encased in bandages, he walked back to the police car. Joe believed in his ability to talk himself out of traffic tickets and was usually successful. I knew his luck had run out this time and was thinking it will teach him a needed lesson. He was gone a really long time. When he returned, he was grinning.

"Don't tell me you talked your way out of this ticket!"

"Yep. Guess what? He's going to join Escapees."

The park, built exclusively for Escapees members, would be run by the 120 families who bought a lifetime membership in the park. They elected their own volunteer board of directors, set the park rules, added improvements through fund-raisers, and maintained the roads through an annual fee. The only rules we imposed were that they provide rental spaces to traveling members, they could not sell their membership for more than what they had paid, and all residents must remain members of the Escapees Club. This last rule was impossible to enforce, so we hoped people would respect that these parks were built by Escapees *volunteers* for the benefit of other Escapees members.

With the purchase of a membership came entitlement to a specified lot in that park. They were free to turn in their membership at any time and the

Escapee buying their membership gave them the same amount they had paid plus the cost of permanent improvements such as a storage building. That rule meant people were not tied forever to their lot, but it didn't allow the membership to be used as a money-making investment. We advertised for them, helped get volunteers and members, and advised them on legalities.

Casa Grande, Arizona was our first SKP Co-Op Park.

Even before the final touches to this first SKP Co-Op were added in 1982, Joe and I went looking for affordable land for a second home-base park. Either land cost was too high or water rights could not be obtained. After searching as far east as Mississippi, we gave up and headed to California to replenish our own savings.

On the way, we stopped in Carlsbad, New Mexico, to visit Joe's former school buddies, Wayne and Faye Jean. During the exchange of stories, Joe told Wayne about the park we had built and that we were unable to find a place to build a second park.

"I have a 6,000-acre cattle ranch with the water rights. Maybe it's what you're looking for. I can take you up in my plane and show you."

After they returned, I saw Joe's excitement. "This is exactly what we're looking for. Wayne says we can pick any 15 acres we want. We'll drive up in the morning and walk the land, so we can pick the spot we want."

The next day, it seemed as if the four of us walked miles, trying to find the most suitable place with road access. We decided on the only piece of land that had a few trees. It was in Lakewood, halfway between Carlsbad and Artesia where our daughter, Janet, lived. We purchased the land.

After visiting Janet, we headed again to California, parked in Scott's yard, and Joe worked out of his home local until time for our fifth annual Escapade the following March. Members interested in a second home-base park sent their money and an attorney completed the documents.

After the Escapade, we led a small band of eager volunteers to the place we now called "The Ranch." During the spring of 1983, Joe supervised building this second home-base park. We had all the hookups installed in four months because New Mexico building laws were less stringent than in Arizona and the land more cooperative. At the end of June, those who bought a Ranch membership made their lot selection. It was decided to wait and build the clubhouse in the fall. Many people wanted to enjoy their summer travels, and the ones who stayed wanted to begin improving their selected lot.

At the same time, other things were happening. Members wanted local chapters so the people in an area could get together more often than at Escapades. The first chapter was formed during the 1984 Escapade and named the Golden Gate Chapter because of its location near San Francisco.

Then Joe announced he was turning the organization and building of all future SKP Co-Ops over to chapters. More chapters began forming across the country. Many were formed to build a home-base park in their area but continued as chapters after the park was built.

There was a bigger need than we anticipated. Many full-time RVers wanted the security of a guaranteed place to return to if one of them became sick, died, or they wanted to rest for a while.

Joe wanted to start tramping and earning money again. We had raised the club dues several times, but even with an increasing membership, we were dipping into our own savings. I felt as if we had grabbed a tiger by the tail and couldn't let go.

Yet, it was satisfying to see the results of both parks Joe supervised and realize that, through the efforts of a small group of volunteers, useless land was transformed into a home base for 120 families in two different states. Volunteering was—and still is—the heartbeat of Escapees, and the philosophy of caring and sharing is the reason for its continued success.

While we were building the first park in Arizona, Cathie and her first husband continued traveling. In their early years they did whatever jobs they could find; later both became millwrights working on nuclear and coal power plants. Then in 1982, Cathie was called to Texas to help take care of a close friend (Escapee #10) who had been severely injured in a car crash. She started handling our book orders and helped answer mail.

It was still early 1984 and Escapees had grown to an extent that we needed a telephone so potential members could join or ask questions. (Cell phones existed but were not popular due to the expense.) While helping her friends, Cathie fell in love with Texas, so when we started looking for land on which to put our headquarters, she suggested the Livingston area of East Texas.

It was a logical choice because it was close to I-10, an east-west major freeway, plus both a north-south interstate and two state highways with access to many towns. Texas had other advantages, including reasonable cost of land, less stringent building codes, and no state income taxes. Still we could not afford to buy the land we needed on our own.

Just a few years earlier when we started building SKP Co-Op parks, we were able to use the members' money to build with. Then the federal government passed laws forbidding it. That brought building further SKP Co-Op parks to an abrupt end. The only way to continue was to change to Escapees-owned parks. Because Escapees was still barely getting by financially, we sold some of the land in half-acre lots to members who wanted a Texas home base.

Instead of a membership, the new buyers *owned* their land outright. This eliminated restrictions some SKP Co-Op Park members were arguing about in regard to size and type of homes they could have. Members who owned their land could have whatever size home they wanted.

Solving one problem created others. Some people now bought the land to use as an investment rather than its intended purpose. In time, the "investors" more than tripled their money; but in doing so, they caused everyone else to pay higher property taxes.

Another mistake we made was that Escapees has no funds, except camping fees, to maintain roads, clubhouse, laundry, restrooms, and showers in these new home-base parks. Had we looked at the future clearly, we would have included a resident's maintenance fee like SKP Co-Ops have.

Our first concern was finding the best land in Texas for our headquarters. In April 1984, while on the way to the Escapade, Cathie and her husband found what seemed perfect land in Livingston. After the Escapade ended, the four of us headed to Texas to see this property.

It was primitive land covered with small trees and bushes. The hardwood trees and bigger pines had been harvested leaving behind tree limbs buried in dense shrubbery that had grown over it. A single lane logging trail led into the heart of the land. A lot of work was required before we could build anything. Still, we purchased the first 13 acres of what would eventually become a home base for more than 200 families plus two campgrounds for traveling members.

The land was so densely covered with shrubbery that we could not park on it, so we asked the family living in a mobile home across the street if we could temporarily park our trailer in their yard while we cleared our new land. We hooked into their water and electric for a reasonable fee and used a Blue Boy (blue dumping tank on wheels) to dispose of waste water and sewerage. From April to July, Joe and I, plus occasional volunteers who dropped by for a few days, cleared enough brush to allow rigs to drive to a single electric pole the power company installed.

On July 4, a dozen rigs parked around that electric pole for our first official birthday barbecue. This included a lottery drawing for lots for those who purchased them. Cathie and her husband joined the town's baseball team. When team member Bud Carr, a real Texan, heard that the big barbecue "those Yankees" were putting on would consist of hamburgers and hot dogs, he offered to cook a "Real Texas Barbecue" with ribs, brisket, sausages, and chicken.

It marked the official establishment of a national headquarters. After the birthday rally, nobody wanted to stay and work in the Texas summer heat and humidity. Bob Gambol offered to stay and act as host. Electricity, plus being a dedicated boondocker, made it possible. Our neighbor allowed Bob to get water from their outside faucet.

Joe and I headed to California to check on the kids who were there and work to rebuild our savings. We discovered a new way to cut expenses. One Escapees couple were teachers who wanted to travel for the summer and asked us to "yard-sit" to discourage vandals. It was a happy exchange of services. Our being there protected their property; we had free rent plus the use of their swimming pool, Jacuzzi, and garden vegetables.

When summer ended, Joe announced in our newsletter that we were going to have a work-rally in October to finish clearing space for a 13-site campground plus a clubhouse. Many responded. It was the first Oktoberfest and became an annual event put on by Escapees staff.

Joe was back in the business of building another park, but this one was different. It would also be our personal home base between travels. Even with Cathie's help, we found ourselves consumed with club duties. We never tramped again after 1984 but continued with annual trips by RV and bus tours.

The national headquarters building later became the clubhouse.

During her 11 years of full-time RVing and tramping, Cathie owned two RVs. The first she called "The Rainbow." In 1979, with both earning good money as millwrights, they traded it for an Avion trailer that she named "Rainbow Too." Now parked at our headquarters, and their marriage crumbling, it looked as if her 11 years of traveling were over. She wanted to name our first Escapees-owned park Rainbow's End. It was a good name for a home-base park for RVers.

By 1985, Cathie's marriage was ending and she became our permanent administrator. That summer she started the mail-forwarding service that members had been requesting. At first it consisted of an old-fashioned desk with pigeonhole slots in which she filed mail alphabetically.

She took time to attach a note to each forwarded envelope because we knew most members and were aware of who was sick, who had been traveling, and who had a special event happening. Sometimes we mourn for those "good old days," even while realizing there are many advantages to having a larger membership.

In 1986, our chapters had grown to 19, and we needed to hire directors to coordinate chapter events and help them grow. Although others helped start chapters, the first "official" chapter directors were Dan and Adele Clifton, followed by Denny and Susie Orr.

In 1989, we appointed 20 people to be our first advisory council. We would meet once a year to discuss problems they heard members talking about. Their job was to correct erroneous information members had and meet as a group to help us decide the best way to provide services that fit the members' needs.

We chose council members who were dedicated to Escapees concepts, active in the club, and had prior experience in an area we felt was helpful. Except for the first group, they served for three years. Those in the first group were appointed for one, two, or three years so their replacements would be staggered. These people had confidential information about our future plans for new services, so they also had to be trustworthy.

The first council meeting was held in Livingston in October 1989. The night before the meeting, tropical storm Jerry was upgraded to a hurricane and was aimed at Houston. Since Livingston is over 70 miles north and inland, we didn't get the full blast, but it spawned tornadoes and high winds.

The storm awakened us; we had no electricity, but that is a common occurrence. In the morning, we got in the car and started out the driveway when Joe slammed on the brakes. My favorite (huge) holly tree had fallen across the road. We would have to walk the half mile to the clubhouse where everyone met to pool rides.

We passed lots of fallen limbs but nothing serious until we got to the land bridge crossing a deep ravine. Another fallen tree completely blocked it. We climbed over it. At the clubhouse, we hitched a ride to the meeting held at a local motel with a restaurant and no electricity. Not a great way to begin a new program, but Escapees are the kind of people who have learned to be flexible.

At the end of a long day, we returned home to find someone had cut up both trees, making the road passable again. That is the kind of thing Escapees do for each other without being asked. We never learned who did it.

Escapees has its own post office and separate zip code.

The mail service was growing. We built a second building and moved the mail service as well as the office staff to it in 1991, using the old building as a clubhouse to expand park activities. It seemed that every couple of years after that we added onto the national office building, including a second floor.

The mail service took up an enormous amount of space and had its own staff to answer phones and send out the mail requested for that day. It became a small post office with its own zip code. The U.S. Postal Service delivered and picked up mail daily in an 18-wheel truck. The photo shows only one section of the mail room which takes up most of the first floor and has serviced over 12,000 families.

Some members wanted us to close the membership when it reached 1,000 members. How could we justify that when new people needed the support, knowledge, and parking we provided?

Cathie and Bud Carr with "best man" Greg Carr.

In September 1987, Cathie and Bud Carr were married at the clubhouse. It was the first of several marriages that took place at Rainbow's End. Theirs was a traditional wedding since Cathie's first marriage was by a county judge. Bud's son, Greg, was the best man, and Joe proudly walked Cathie down the aisle.

For the first decade, Joe and I ran the club with volunteers and a few paid staff. By 1990, we were tired of not having any private lives. We wanted to do more traveling while we had the health and energy. We were also committed to many speaking engagements across the country, so we turned the running of the club completely over to Cathie and Bud.

They took over at a time when RVers were being challenged about their right to vote. It was our policy not to take sides on religious or political issues. Now we were drawn into political battles after the industry reported there were hundreds of thousands of full-time RVers. Questions like where to register vehicles and whether "nomadic people" should be allowed to vote became hot issues that threatened the rights of full-time RVers.

We needed to educate lawmakers who thought we threatened their security. No one in the RV industry seemed willing to get involved, so Escapees stepped forward. Bud and Cathie Carr led the fight in re-establishing the right for Texas RVers to vote.

In winning that fight, they educated many people about the RV lifestyle. Shortly afterwards, Escapees became emerged in a fight with some campground owners who saw boondocking (free parking) as a threat to their income. Again, education and compromise were the answer. Escapees established an overnight parking creed, called "Good Neighbor Policy," that pacified most campground owners. It took an entire decade to obtain the support of 14 of the most highly respected RV clubs in the country.

While this was taking place in the background, Escapees continued adding services and educational programs to make RV life easier. The rapidly growing RV industry took note of our accomplishments; we received many awards and recognition during the 2000 decade.

In August 2001, Joe and I were the first *couple* to be inducted in the RV/MH Hall of Fame in Elkhart, Indiana. Unfortunately, their plaque recognized us as campground owners and said nothing about our founding of Escapees.

In 2002, Mark Nemeth became our technical advisor and helped establish our popular Boot Camp educational program.

In 2003, Escapees received the award from Baylor University for a Texas family business.

In 2005, Escapees magazine won a national media award under editor, Janice Lasko.

In June 2006, I won the Polk County, Texas, senior-of-the-year award, and in November, Escapees won Baylor University Texas family business award for the second time.

In June 2010, Joe and I were named by the RV industry to be among 100 of the most influential people of the past *century*.

By 2008, with Escapees now 30 years old, the third generation of my family introduced new technology ideas and innovations. The country's economy was a disaster. Along with other industries, RV businesses were on a downward spiral. Many companies closed their doors on the verge of bankruptcy. My granddaughter, Angie Carr, Escapees executive director, found ways to keep Escapees financially solid by automating labor-intensive tasks and controlling expenses through streamlining old programs and introducing new ones.

Some thought RVing would become a past luxury. I knew better. Even though we personally gave up our RV because of Joe's eyesight and my inability to drive, I believed there would always be too many devoted travelers to let RVing die. The ones who sold their RVs were vacationers.

Full-time RVers made changes in their travels and many downsized because of gasoline prices, but the RV lifestyle is still a less expensive way to live. RVs are sold every day to people whose once-secure job has vanished or their luxurious lifestyle is impossible to maintain. I believe there will always be RVers, and Escapees is the support network they need.

The Carr family from left around table
Bud (2nd gen,),Baby Gabe (4th gen.), Melanie and Travis(3rd gen.)
Cathie (2nd gen.), and Angie and Greg (3rd gen.)

Today our original 4-person board of directors consists of six directors plus me as a consultant director and tie-breaker if needed. The picture here was taken at their 2012 annual planning session, which I was unable to attend. However, you can see my great grandson, Gabe, attentively listening. I hope he will be the fourth generation for this family-run organization.

When Joe and I started the club in 1978, our purpose was not to start a business, make money, or become famous. Our goal was to help like-minded people achieve their RVing dream. Full-time RVing was the best part of our lives, and we wanted to share it.

Today's technology and the scores of books on advice from people who are currently doing it, make full-time RVing a cinch. However, nothing will ever replace Escapees, Inc. as the best support system for all RVers. The future looks brighter than ever.

(For more information, check www.Escapees.com.)

CHAPTER 15

Kay Peterson's America (1973—1991)

WHILE TIM WAS still with us, we spent one Christmas in Death Valley, a place with a desolate reputation. Even in December, Death Valley is so hot that desert critters hide during the day. No sane person would be hiking up and down those hills, and yet in front of us a young man with a bedroll on his back trudged up the hill. We stopped to give him a ride to the campground. It was the only humane thing to do.

His name was Veerof, and he was an exchange student from India studying at San Francisco University. He used the holiday to see more of California. He hoped to get to Las Vegas before he had to start back. Joe and I looked at each other, passing a silent signal to include Veerof in our Christmas celebration.

After setting up camp, we all walked to the castle. After I purchased a few stocking gifts, we walked back to camp. Veerof happily accepted our invitation to supper and to help us decorate a small Christmas tree. We talked about the different customs of our countries. When Veerof retired, Tim hung his traditional stocking. I found a stocking for Veerof.

In the morning, Tim called, "Veerof, come see what Santa brought." When Veerof saw the stocking with his name pinned to it, his eyes lit with childlike enthusiasm. Mimicking Tim, he pulled out the assortment of fruits and candies, exclaiming over each one. When he came to a small package tucked into the toe, he seemed overwhelmed. It was a $3.50 collapsible drinking cup imprinted with the emblem of Death Valley.

"I want to make you a gift also," he said. "It was in my mind to give it to you when we parted, but now is a better time." He hurried outside and came back with a brown paper sack which he handed to me. Inside, wrapped in newspaper to protect it, was a vase I had admired in the store but didn't buy because I had enough "stuff." I don't know when Veerof bought it.

I no longer categorized the vase as "stuff." It was on display in every RV we had and still has its place of honor. I wonder if Veerof treasures his souvenir of Christmas with an American family the way I do mine. I like to think he does, but if he doesn't, it does not diminish the value of my vase. The cup and vase are symbols to jog our memories of the magic of Christmas.

Some other experiences stand out like photographs engraved on the pages of memory. When I see shrimp, I think of the fisherman we met one Sunday afternoon in Freeport on the Gulf Coast of Texas. Freeport claims to be "The Shrimp Capital of the World." Since we had never seen a shrimp boat, we decided to stop there.

After parking out of the way, we walked to the waterfront where the Golden Dawn was undergoing a beauty treatment. We watched two men painting her rigging. One called a greeting. I called back, "This is the first shrimp boat we've ever seen."

"Y'all come aboard, honey. I'll have Captain Ray show ya 'round," he called back.

Captain Ray turned out to be a Cajun, tall, well-built, with a weathered brown face. He told us he had been a shrimp fisherman for over 20 years. "Before that, I was an iron worker," he said. "It paid more, but I like the sea and being my own boss."

After he gave us a tour of the boat, we sat on deck sipping from beer cans. "We're fixin' to enter the 'Blessing of the Fleet' parade next Sunday," he said. This religious ritual dates back to time immemorial. Fishermen believe if their boats are blessed, it helps them have a good catch and a safe return. In Freeport, it is an annual April event.

Lawrence, the man who invited us aboard, said, "We're fixin' to work to midnight every day 'til we get 'er decorated."

The captain explained, "We string colored pennants along the mast and lines and then decorate the whole boat with crepe paper flowers."

"It takes lotsa flowers," Lawrence added.

In 1957 the ceremony was broadened to include a parade of decorated boats down the Brazos River. It was discontinued for a few years and then revived in 1973. Now it was an annual festival with the religious ceremony part of it.

"Y'all oughtta come back next weekend and see for ya' self," said Captain Ray. "Them decorated boats, with ensigns flyin' in the wind are a glory sight."

"They give money for the best-decorated boats," interrupted Lawrence, "and we're fixin' to win first prize!"

"Friends can ride on the boat. Y'all wanna be my guests?" the Captain asked.

"We're fixin' to have a weddin' on the boat," Lawrence added.

That did it! We assured them we'd be back the following Saturday. Captain Ray told us to park our trailer in the driveway of the fish-packing house where Golden Dawn was anchored. Dan Allen, owner of the boat, was also manager of the fish house.

The following Saturday, when we pulled into the driveway, Dan came to greet us. He was a slender, wiry man of average height, probably younger than he looked, with a smile that lit up his face. He was dressed in working clothes and a baseball cap. It was hard to imagine him owning a boat. He introduced us to his wife and the daughter who married Captain Ray. They were serving a shrimp, fish, and bean supper to friends and employees. Dan insisted we join them. While we were eating, someone came in to report a storm headed for Freeport.

"Does that mean you may have to postpone the festival?" I asked.

"Don't worry," Dan said. "It never rains when we have the parade. We've never postponed a festival."

I noticed none of the boats were decorated except for colored pennants made of durable plastic. I saw cardboard boxes, filled with paper flowers, stacked in a corner of the fish house. Dan said, "We wait 'til the last minute to decorate. A heavy rain'll wreck paper flowers."

The coast of Texas is a prime target for nature's vengeance. Throughout its history, the Brazos port area, which consists of a few villages and seven incorporated towns including Freeport, has been abused by sudden freezes, unexpected storms, and tropical hurricanes. A less determined people would have packed up and moved out, but shrimpers and farmers along the Gulf Coast are a stubborn breed. They rebuild homes and repair the damage after each disaster as if they were sure it was the last one.

So even though Dan expressed confidence Sunday would be as sunny as those of past festivals, he did not tempt the gods. We were invited to help with decorating the next morning. After supper, Dan said, "They're fixin' to have a big street dance tonight. Y'all oughtta go. Y'all kin use my car so's you won't havta unhitch your trailer."

We declined his offer but wondered about a man who would loan his car to people he had only known a few hours.

We were getting ready for bed when the wind, like a youth flexing his muscles, boasted its strength. That storm was one of the most violent I've seen. As I stared into darkness, I felt as anxious as the first night long ago when I waited for Louise to return from her date. When lightning flashed, I could see the wind whipping the Golden Dawn's plastic pennants as they thrashed in angry protest. Sometime during the night, the wind moved on taking the rain with it.

Sunday morning pushed its way through an overcast sky. When Dan and his family arrived shortly after dawn, we went down to the boat to help them sop up the water and attach the decorations. Lawrence was right; it takes a lot of flowers to decorate a boat. All along the waterfront, other boats were being decorated in like manner. I wondered how many hours of labor had gone into making all the paper flowers.

After our boat was decorated, Joe and I drove to town to attend the bazaar and other festivities. By now, the sun had burned a hole through the clouds, and the air was electrified with excitement. Stepping around puddles, we made our way to a seafood contest. We saw shrimp that had been prepared in every imaginable way. It was served ice cold or steaming hot in a variety of tangy sauces. There were shrimp salads, shrimp casseroles, shrimp stroganoff, and a variety of other dishes.

In addition to food, every item imaginable was sold at the booths that lined the boardwalk. There was a beauty contest for teen-agers to compete for "Queen of the Shrimp Festival." At noon, everyone gathered around a roped-off area where a drill team of 13 black prisoners from Retrieve Farm put on an outstanding demonstration.

When we returned to the Golden Dawn, the band Dan hired was setting up their instruments. Our boat was filling with invited guests. (I felt as if the Golden Dawn truly was our boat.) We were number 12 in the parade, so it would be some time before our turn came to line up in the river.

The arrival of the bride and groom created a new wave of excitement. The groom was also the base guitar player. Both he and his bride had been married before, so eight children watched them exchange vows. It reminded me of our marriage. When the ceremony ended, it was a signal for the party to begin.

Someone opened a keg of beer, and people served themselves from a huge pot of shrimp gumbo made from a secret recipe. The band started playing. Dan, still in his baseball cap, started dancing. Little children to grandparents joined in, their bodies twisting to the haunting beat. Most were dancing singly, not as couples.

As the music and gumbo aroma mingled and drifted around me, it seemed some were in a trance. They not only heard the music, they felt its sensuous touch.

What a joy watching their rhythmic bodies! Many never stopped dancing.

When it was Golden Dawn's turn to move out onto the river and start its slow voyage past the people-lined shores, those of us not dancing waved to the cheering crowd. We passed by boats that were not decorated. Since owners paid their own decorating expenses, some didn't have the money and couldn't find sponsors, so they couldn't participate in the parade but would go at the end to have their boats blessed.

A priest and ministers from three churches stood onboard the lead boat in the parade, which was now anchored at the far end of the "run." Each fishing vessel circled past the lead boat and was blessed. This ritual is important to fishermen who are exposed to many perils at sea as they search for fast-disappearing shrimp.

It is not an easy life for them or their prey. During shrimp spawning, which takes place along the 375-mile Texas coastline, numerous marine animals, including larger shrimp, feed on the newborn eggs. In the warm water, survivors grow rapidly, shedding their outer shells layer by layer as they grow. While they are growing, they are besieged by predators.

Fishermen were the greatest predators of all. Yet life isn't easy for them either. In the past years, they had to cast their nets to catch the shrimp. Then, around the beginning of World War I, a trawl was designed to replace that back-breaking labor. Yet, even with modern improvements, shrimping is difficult. Oil spills may someday extinguish both shrimp and fishermen. (I wonder what has become of Golden Dawn.)

Shrimping is usually done on a shares system. Half the profit goes to the owner of the boat, and the other half is divided among the crew. On most boats, the crew consists of the captain and two hands, but on smaller boats a husband and wife often work as a team. "Many times we go out and don't even make expenses," said someone who started shrimping at age 16.

"It's dangerous, too," chimed in Lawrence. "Oneced, I nearly died." Lawrence explained that on one fateful trip he and the captain went to bed, leaving a rookie in the wheelhouse. Since they were dragging at the time, there wasn't much to do. The rookie was watching the other boats and didn't notice the bilge pumps had quit working until the wheelhouse started filling with water.

"Too late," Lawrence explained. "No way ta save the boat, so we went over the side and swam all the way to shore. Man, that water was cold. Me 'n' the captain made it. Rookie drowned. That was God tellin' me was time ta stop shrimpin' so I works in the fish house now." For a long moment, no one spoke. "Sometimes I help with the boats," Lawrence added, "but I ain't goin' to no damn sea again."

Many fishermen exaggerate a story to make it more interesting. I suspect Lawrence did because it would make more sense to swim to another boat than all the way to shore. Most fishermen are superstitious. That's why it is important to them to have a sunny day for the Blessing of the Fleet.

When the Golden Dawn approached the bridge in front of the anchored boat, the band stopped playing. Those who wore hats removed them, and everyone stood in solemn silence as we turned and passed in front of the boat in which the clergymen performed their ritual. Above them were the judges who decided which boats should win the three cash prizes.

We cruised down the river, making a second sweep. As we passed by other boats in the parade, I realized ours was the only boat with a live band, but it was obvious some were more elaborately decorated than we were. However, I didn't see anyone who looked as if they were having as much fun as the people aboard the Golden Dawn.

When the parade was over, there still remained a contest for best-dressed captain, the awarding of various prizes, and a ceremony to crown the Shrimp King. This last honor was given to the owner who had made that year's largest catch.

While Dan and his family attended these events, the Golden Dawn was tied to another boat. The band continued to play, and those on shore were invited to come aboard and dance. And come they did. It was another three hours before Dan and his family returned with news that our boat did not win any cash but took fourth place for the band.

The festival was over for another year. It was time for Golden Dawn to head back to the fish house. Captain Ray announced we were leaving so those who wanted to could get off. The announcement was repeated several times before the boat eased out into the river. There was laughing and waving of arms as those aboard tried to communicate with their friends on shore.

"All y'all come get us at the dock." Some would end up walking two miles back to town. One man took off his shoes and his shirt, gave them to a friend, climbed the rigging, dove into the water and started swimming back to shore.

Night was settling and people were still dancing when we docked. Then the band packed up their instruments, but we were told the party would last as long as anyone wanted to stay.

As Joe and I started up the ramp toward our trailer, we passed a man carrying two cases of beer toward the boat. Dan stood at the end of the dock saying good-bye to those who were leaving. He was still wearing his baseball cap. He threw his arms around me. "Y'all come back next year, y'hear." He grinned. "We're fixin' to win first prize next time."

"Dan, as far as we're concerned, you already won first prize."

You can only have serendipitous adventures if you are curious about what's happening around you and willing to change your schedule when an opportunity is presented.

It was late afternoon on I-25 in eastern Wyoming when the sky darkened and God turned on a faucet above us. As we took the off-ramp marked CHUGWATER, we saw a homemade covered wagon being pulled by three mules and a horse that must have seen better days. A fourth mule, which was probably their spare tire, was tied to the back of the wagon. A dog sat between two young men occupying the only visible seat. The entire entourage looked as if it belonged in a Hollywood Western.

We let them go in front of us. When the mules reared in protest, the primitive wagon came to a squealing halt in front of a cattle guard on the road. Fences on both sides indicated there was no way around it. Pulling onto the shoulder of the road, Joe and I defied the rain and went to see if we could help. Within the hour, we managed to round up enough scraps of plywood to cover part of the cattle guard and convince Jack and Number Seven, the two lead mules, to walk across this makeshift bridge.

After getting the animals across, we looked at each other and grinned. By now we were drenched, so we agreed to find a place where we could camp with our new friends. It took vigorous coaxing to get Jack and Number Seven to lead the way into Chugwater, which consisted of two gas stations and a store that was closing for the holiday weekend. We asked permission to park in the empty lot in front of the store where Jack had already stopped.

Over the past two months in his position as lead mule, Jack clearly established he had a will of iron. Coaxing, threatening, or beating with a stick would not convince him to do anything he didn't want to do. When he reached that empty lot, Jack refused to pull the wagon one inch further.

After picketing the animals where they could graze, Pat and Carson changed into dry clothes, left the dog to guard the wagon, and joined us in our trailer for a hot meal and to share the details of their 6,000-mile trip in a covered wagon.

"I started from Tulsa, Oklahoma, in September of 1976," Pat said. Pat, who was over six feet, was 22 years old. His love of horses began when he was a child. No one else in his family liked them, so Pat saved his money until he had $45, enough to buy his first horse.

All he knew about horses was that you could ride them. He soon discovered riding one isn't as easy as it looks in the movies. When his father, a doctor, set Pat's broken bones the *third* time, he told Pat if he broke one more bone, the horse would go. So when Pat broke a couple of fingers and cracked some ribs, he told no one.

When he was 17, Pat went to Farrier's College to learn horseshoeing because that service was expensive. After he completed the one-year course, Pat joined the army. While overseas, he decided he would tour the U.S. when he was discharged. Only he would do it by "trekking," which means on horseback or with a covered wagon.

Pat planned to go from his home in Oklahoma to Arizona, then Canada, and then Texas. The pioneers did it, so he was sure he would have no problem with modern equipment, decent roads, and towns along the way to get water. He didn't even have to worry about losing his scalp to Indians.

After being discharged, he bought a pack mule and April, a registered black-and-white paint. While making preparations to leave, a dog adopted him. "I named him Mac because he looked like that should be his name." With $400 in his pocket, and Mac trotting beside him, Pat mounted his horse and headed for Arizona, where he planned to spend the first winter.

Soon the money was gone. Some went to buy new gear after Pat lost his camera, sleeping bag, and most of his food. "Didn't exactly lose them," Pat admitted. "I threw them away."

While crossing Oklahoma, he spent the night in a farmer's barn. The next morning he had already saddled April when the farmer invited him to breakfast. He left April untied so she could graze freely. They were eating when the farmer's son, who was facing the window, exclaimed, "Hey, cowboy! Your horse is going down!"

He looked and thought he saw April sitting on the grass. Something must be wrong. She always stood with her saddle on. Then PLOOP! She was gone. He watched in open-mouthed amazement as she came up and then disappeared *three* times.

"We'll never get her," the farmer said. "That cesspool is 25 feet deep."

April didn't know that. She continued to struggle until she got her two front feet on solid ground. Then she rested her muck-covered head on the bank and waited patiently for Pat and the farmer to get a tow rope and a pickup truck to pull her out.

"It took three gallons of shampoo to clean her," Pat said. "I saved the saddle, but didn't use it for a while. For a week, I walked downwind of her."

Pat admitted that most of the $400 was spent foolishly. During the first weeks, he stayed in motels and ate in restaurants too often. With winter dogging his heels, he was no further than Lipscomb, Texas, and flat broke.

Lipscomb is not exactly a metropolis. It has a combination store-café-gas station and a post office to serve the town's 30 residents. Harry, one of those residents, owned a ranch and hired Pat to break horses. While working there, Pat met Harry's nephew, Carson, who was intrigued with Pat's trekking plans. He watched enviously when Pat left in March.

With the money he had earned at the ranch for a grubstake, Pat got as far as Colorado before his trip was interrupted again. While passing through New Mexico, Pat met another trekker headed in the same direction. They agreed to travel together.

Pat wasn't wasting money in restaurants and since beans alone make drab eating, he shot a rabbit or a pigeon to add to his meal. His traveling companion thought that was great and asked to borrow Pat's rifle to go rabbit hunting.

"By the time I figured out he wasn't coming back," Pat explained, "he had a pretty fair start." Pat was furious and wanted his rifle back. For three days he rode hard from sunup till dark chasing the thief. "Finally got over my hissy fit and thought how stupid I was," Pat acknowledged. "I was riding' my horse into the grave to get back a $15 rifle."

It was too late. April was saddle-sore. The tree of the saddle had broken and bruised the horse's shoulder. Called a "sit-fast," this type of sore cannot be seen until it works out to the surface. By then, the pain was so bad that if Pat touched her back, April would drop to her knees. Sometimes you can make an instant decision that gives you heartache for the rest of your life.

The first vet said it would take weeks of expensive treatments to cure her; he advised euthanizing her. Pat was not ready to give up. He continued to Denver, the logical place to find medical care and a job to pay for it. He found both at the racetrack. While the track vet treated April, Pat worked as security guard and moonlighted by shoeing horses. By the time the track closed down, he had spent over $600 in vet fees, and April was no better.

Completely discouraged, Pat was trying to decide what to do when Harry arrived at the track to buy some horses. When Harry heard Pat's story, he offered to take the mare to his ranch where she might recuperate out in the pasture. Pat spent all his earnings on vet fees and no longer had a horse to ride, so he went back to work for Harry.

That winter, working with Carson, they became fast friends. Together they built a covered wagon, improvising when they couldn't find the needed materials. When spring came, the journey would continue, and Carson would go with him.

Harry wanted a colt out of April, who was now recovered, so Pat left her and traded her unborn colt and his old pack mule for two pulling mules and a black stud horse. In late March of 1978, the two men and the ever-faithful Mac set out in the covered wagon. They were barely out of Texas when one of the mules broke a leg and had to be shot. The stud horse caused trouble whenever he got close to a mare, so they traded him and the second mule for four new mules and a "bucking bronco" that turned out to be a gentle old mare.

Trading was a big part of Pat's life. He trades the jewelry he and Carson make out of arrowheads picked up on the ranch; he trades his horse-shoeing expertise; in fact, Pat will trade anybody for most anything.

Proof came while we were eating supper. Whenever Pat arrives in a town, he asks the curious onlookers if they know of a horse needing shoeing. Today was no exception, and word spread. We had just finished supper when a rancher came to bargain.

Iona was a crusty little woman with an invalid husband. She ran their ranch and wanted to be certain Pat knew what he was doing. An inexperienced farrier can cripple a horse. After Pat convinced her, the haggling began. For more than an hour they made offers and counter offers until they agreed Pat would shoe one horse for a used trace chain link, a Coleman lantern, a piece of old plywood to use on the next cattle guard, a gallon of milk, and $10 in cash.

"We'd never make it if we didn't work as we go," Carson confided. So far, they had worked or traded for the grain and other supplies needed. "We cut down to one meal a day and coffee. Now it's cold beans and dried antelope meat 'less we get lucky 'n shoot a rabbit."

Their biggest expense is the animals. It took $6 a day for grain to feed five animals. When they can, they supplement this by letting them graze. In Cheyenne, they spent the night in front of a drive-in theatre where the animals could graze.

"If there wasn't no place to graze the animals, when the lookie-loos show up, Pat asks if anybody knows where we can get hay. Folks will start calling friends, and usually a farmer shows up with a bale. Some farmers wouldn't take any money," Carson said.

I asked if they ever encountered a problem with the police because of their unusual rig or the unorthodox places they spent the night. Before they started, they wrote to the governor of each state they expected to pass through. All sent copies of their rules of the road, and a few answered personally.

"The governor of Colorado encouraged us to come through his state and invited us to visit him in the capitol. We been places where local ordinance forbids animals, but no cop cited us yet. Our biggest problem has been getting water to fill the two 55-gallon wooden barrels. Animals have to drink huge amounts of water a day."

There would be other problems in the future. The next leg of their trip would extend into winter. "I'll be fine 'cause I got my bag," Carson said. "Pat only has blankets."

"Why didn't you replace the sleeping bag that fell into the cesspool?" I asked.

"I bought another one but had to throw it away in New Mexico," Pat said. "It got cold one night. When I woke up the next day, I felt something on my belly. Looked down and saw a rattlesnake curled up sleeping there. Jumped outta the bag, grabbed my forty-five, and blew three holes through that snake. I shoulda knocked the snake off the bag before I shot. Fact is, when you wake up and find a rattlesnake sleeping on you, you get a tad excited."

Problems do not discourage Pat and Carson. When we left them, they were excited, hoping to make it to Canada in time for the chuck wagon races at the Calgary Stampede. "If we don't make it this year, we'll try again next year."

They were in no hurry to finish the 6,000-mile trek. They knew it is not the destination that matters but the trip itself.

Serendipitous meetings are great, but being a writer for a magazine gave me additional opportunities. I was able to interview two of the major trainers at Sea World in California. The first was Ray Keyes, curator of fishes, who was responsible for the health of *all* the marine life at California Sea World. If a shark was sick, Ray gave it medicine. Most of us have tried giving medicine to an objecting child, but how about an objecting shark?

"The only way to get medicine into a shark is by force feeding," Ray explained. "It takes two people. One holds the shark while the other inserts a feeding tube down its throat to get the medicine directly into its stomach. The shark isn't happy about it. When I started working here I learned sharks are hand fed, and it was my job to feed them."

Ray explained that there are more myths about sharks, and less actual knowledge than any other creature. "Sharks can be *dangerous* but will not bite you—unless you move when he bumps into you. It takes self-discipline to remain perfectly still. Employees only go into the shark tank to feed them or give them medicine. Sharks don't like human contact, but I love working with them because they are highly intelligent."

At Sea World, the killer whale show is the most popular. The leading performing whale uses the stage name of Shamu. Next I interviewed Bruce who started working at Sea World while a student at San Diego State College. He was majoring in behavioral psychology with the intention of being a psychologist but needed to work while he finished school. He answered an ad in the local paper for a high-wire diver with Sea World's productions department.

One of his duties was to take part in the dolphin lagoon show where he did "insane things like diving off a high wire strung 30 feet above the water." He became enraptured with dolphins and decided working with these wonderful creatures was his true vocation. He was transferred to the animal-training department. Within five years, he worked his way up to director of training at San Diego Sea World. When I met Bruce, his biggest challenge was the arrival of a new Shamu.

Winston was captured near Puget Sound and brought to a museum in London. When he became too big for the tank they had, he was sold to Sea World. His exact age was unknown, but scientists estimate he was born in late 1962. At birth he was seven feet long and weighed about 350 pounds. Even then his black head with massive jaws and interlocking teeth, plus a vivid black-and-white body, was impressive. He lived in the wild until he was about eight years old, eating tons of fish, birds, and even dolphins who were his cousins.

Killer whales are social animals that belong to a close-knit family and hunt in a pack that is often followed by sharks that eat their leavings. When the killer whale finds smaller animals, like penguins or seals, they swallow them whole. Larger prey, like the blue whale, are attacked by the entire pack and eaten in pieces torn off by the powerful teeth and jaws.

It is the killer whale's voracious appetite, along with his tremendous size, amazing speed, and indiscriminate hunting that caused primitive people to circulate legends about his attacks on fishermen. There is no evidence the killer whale was ever a predator of man. True, when hungry, it eats anything that swims in the sea, making him the undisputed ruler.

When Winston arrived by plane from London, he was the only passenger. Many people are intimidated by his enormous size, but Winston has a true affection for humans. He likes being around them and enjoys being caressed. The San Diego transport crew had no difficulty lowering the water in his tank and climbing in with him. Then they rubbed his entire body with greasy hydrous lanolin to keep his skin moist during the 12-hour flight to California.

He allowed them to place a canvas sling around his body. If he wanted to stop them, he could lean his 700-pound body against them, and their frail human bodies would smash like egg shells against the side of the tank, but Winston had no desire to hurt anyone. There were holes in the sling, and the men helped Winston push his black fore flappers through them. Then, using a crane, they lifted his three-ton body out of the tank that had been home for six years. He had grown to 21 feet.

Once he was settled in the airplane, the transport crew packed ice around him. He was getting warm, so he welcomed this. He relaxed even more when they used hand pumps (used for garden insecticides) to spray his body with salt water.

When he arrived in San Diego, Bruce was waiting. The animals Bruce had worked with until now were young and untrained when they arrived. It was easy for them to adapt natural jumping and twisting skills to perform before an audience. Winston presented a challenge because he had been taught behaviors that could not be used. He had to learn Sea World's way of doing things.

As with humans, breaking old habits is not easy. Bruce's background in behavioral psychology enabled him to use modern theories in animal training rather than outdated methods. Bruce said, "Training animals is based on mutual respect. My methods make it easier to establish a bond. When you withhold food until the correct behavior is done, their minds are constantly on food, and consequently they become aggressive."

Bruce feeds his animals the same amount whether they perform well or not. Winston's diet consists of 140 pounds of squid, mackerel, and herring each day. Killer whales have two basic needs—food and companionship. Bruce believes if you eliminate predatory responses by feeding them well, their mind becomes centered on their need for social contact. This makes both animal and trainer look forward to training sessions.

Although well-fed animals do not attack, there is always a danger of the trainer falling off the whale's smooth-as-glass skin as he zooms through the water. If a trainer fell off, he could be crushed between the whale's body and the side of the tank or could drown if trapped underwater by the huge body.

Bruce rewarding Winston with a treat.

To avoid that catastrophe, Bruce uses systematic desensitization. The whale is taught not to react to outside stimulus, such as a fallen rider, but only respond to commands from the dock trainer. To teach Winston this, one trainer gave commands from the dock while another swam around in a deliberate effort to distract Winston. After months of patient training, Winston learned to ignore the swimmer. Now he could take his turn being Shamu.

If you go to Sea World, look closely when that black and white body shoots out of the water in a soaring leap. If he completes the jump well, you will see a nod of approval from the trainer on dock. Winston will see it too and blow through the hole in the top of his head to show his own satisfaction.

One of my favorite interviews was spending a day with the famous Korczak Ziolkowski, who spent half of his life carving Crazy Horse Mountain in the Black Hills of South Dakota.

It takes a big man to carve a mountain, so I wasn't surprised that he stood well over six feet. He was 70 years old at the time, but still had broad shoulders and muscular arms. His hair, balding under a battered old hat, was as faded as his denim work clothes. His piercing blue eyes, peering at me above a long, wild, gray beard, commanded attention.

He walked slowly in scuffed work boots because his bones complained of constant pain that he attributed to nights spent crouched in a foxhole during World War II. Years of working in the cold and wind on top of the mountain left their mark, too.

When Crazy Horse statue is finished, it will dwarf Mt. Rushmore 25 miles north in the same Black Hills. It will be 563 feet high and "in the round." One of the Mount Rushmore heads could fit into Crazy Horse's hand. A 10-story building would fit inside the space between the horse's mane and the Indian's arm. That outstretched arm is the length of a football field.

Replica of Crazy Horse statue and the mountain status in 1979.

Korczak removed seven million tons of rock before he could begin to carve the outline of a warrior riding a stallion. His project began May 3, 1947, when he used his life savings to buy a granite mountain. He was 39 years old, married to a Massachusetts blueblood, had one small daughter, and no more money.

In the town of Custer, five miles away, people gossiped about this stranger. "He's a crazy Polack from Boston. Thinks he can carve a mountain into a statue."

"Yeah. I heard it's that Sioux chief, Crazy Horse."

"You're kidding? Isn't that the devil that killed General Custer and our troops at Little Big Horn?"

"We got to stop him! It's a disgrace to our town."

"No worry. Nobody can do what he *says* he's going to do."

Standing Bear, an Indian interpreter, didn't agree. They met when Korczak was working on the Mount Rushmore monument under Borglum. After Standing Bear and Korczak became friends, he asked Korczak to make an even bigger monument honoring the Indian people. Standing Bear, a nephew of Crazy Horse, said, "Make us a statue of Tashunca-Uitco so the white man will know Indians have heroes, too."

Korczak accepted the challenge. Although he died long before he could finish the statue, he left detailed instructions, and his family is finishing it for him. It will take many more years and at least another generation of family members to finish it.

The white man thought Indians were lazy. They couldn't understand a whole race of people who rejected possessions, preferred to roam instead of settling down in one place, and saw no need to plan for a rainy day.

The Puritans believed white men were best able to judge how the world should be run, and after much bloodshed and heartache, the white man had his way. One after another of the Indian nations signed treaties to end warfare.

The treaty the Sioux nation signed in 1868 contained the statement: "As long as rivers run and grass grows and trees bear leaves, Paha Sapa (the Black Hills) will be the sacred land of the Indians."

However, six years later, white men discovered there was gold in the Black Hills. Those who wrote the treaty had not anticipated that. They wanted the land back. What did the Indian need with gold?

"They made us many promises," said Red Cloud. "They only kept one. They promised to take our land and they took it."

But not without a struggle. When prospectors swarmed over the hills, slashing away with picks and shovels, the Indians put on their war paint and blood stained the land again. Then came news that stunned Americans. General Custer and all his troops had been wiped out!

The Indian victory was short-lived. Within a year, the white man's superior force and weapons caused Crazy Horse and 2,000 of his warriors to surrender. Four months later, on September 6, 1877, Crazy Horse died of a bayonet wound inflicted by a soldier who "thought" he was trying to escape. Before he died, Crazy Horse told his friends, "I will return to you in stone."

Did he have a premonition of Korczak's mountain? Some people think so. Perhaps Standing Bear did, too. Those who believe in reincarnation point out that on September 6, 1908, exactly 31 years later, a son was born to Joseph and Anna Ziolkowski. They named him Korczak.

There is no other similarity between the Indian's childhood and Korczak who was born in Boston—the most Puritan of all the white man's towns. His parents were killed in a ferry boat accident when he was three years old. He was then adopted by an Irish prizefighter who beat him into submission—something that would never have happened to a Sioux child. Crazy Horse had a happy, loving childhood; for Korczak, the world was a friendless, drudgery-laden place.

When Korczak was 16, he ran away from his cruel guardian and worked at odd jobs while he finished high school. At age 25, he was already recognized as a talented wood-carver who taught himself the art. Then he began the sculpturing career that made him famous.

His friends never understood how he could give up a lucrative career to begin the impossible task of carving such a gigantic statue. It took Gutzon Borglum 14 years to carve four faces at Mt. Rushmore even though he had scores of experts working under him and a million dollars in government funds.

Korczak had no money. He didn't have a horse, but he had a cow he had to learn how to milk. He didn't have water or electricity, but he had a resentful wife and four-year-old daughter to support. For seven months, they lived in a tent while Korczak worked from dawn to dusk cutting down trees to make a passable two-mile road to the highway so he could get electricity and, at the same time, built a dam across a ravine to form a lake to supply their water.

While he was preparing for the future, so was his wife. She took their daughter and left. The people of Custer were delighted. They didn't think any man would spend a winter in a tent by himself. They did not know how stubborn a Polack could be. He began building his house from the trees he had felled.

By the time summer came, people from different states had heard about the project and came to volunteer their help. Among them was a young woman, Ruth, who had lived near him when she was a child and admired his sculpturing. A year later, Korczak married Ruth, and over the years she bore him 10 more children, some of whom still work on the mountain under the detailed instructions Korczak left for them.

I was fascinated as he told us the story of his life while we rode in his jeep up the back side of the mountain to the Indian's out-stretched arm. Then, in spite of his limp, he walked with us the length of this arm to the tip of what would become the Indian's pointing finger.

As we stood looking down on the town of Custer, he told us about the hateful things some townspeople did in an attempt to make him leave. Yet his project was already saving a dying town by making it a tourist attraction.

The most memorable thing he showed us was the tomb he had built into the bottom of the mountain for himself. There was a steel door ready to mount after he died. "Look," he pointed. "There is only one doorknob, and it will be on the *inside* of my tomb."

Every now and then, something reminds me of the day we spent with Korczak and listened to the dream that meant so much to him. He taught me that inside each of us, buried in the clutter of half-forgotten memories, we have our own impossible dream. Then someone said, "You can't do that!" So we stopped trying.

Korczak taught me that you can. And a few years later, I did.

The adventure Joe enjoyed most happened at the Albuquerque Balloon Fiesta, where people from all over the world bring elaborately designed hot air balloons to enter competitions with cash prizes.

We heard the best way to learn about ballooning was to join the pre-dawn spectators who gather at Simms Field launch site. As we walked around watching crews get their balloons ready, I wished there was a way to be more than a spectator. Then I spotted a young woman sipping coffee from a Styrofoam cup. I knew she was connected because her jacket was covered with balloon patches.

"How can my husband and I get to work on a crew?"

Her eyes went immediately to my white hair. "It is heavy work and you have to move fast." She saw my camera. "Will you take a picture of our crew inflating the balloon? I'm supposed to do it, but we're short-handed so I have to help."

Before I could answer, someone yelled, "Come on, Jen. Get your ass over here." She hurried to help her crew spread a multicolored "envelope" on the ground.

I was still snapping pictures from various angles when I heard someone yell, "Hey! Somebody grab that side!"

No one moved. Then, Joe, seeing what needed to be done, jumped forward and grabbed the flapping end of the "skirt." He did not know it, but he had just joined the ground crew. He continued helping them as the balloon filled with air until it was airborne.

I saw Jen talking to Benny, the crew chief. Benny turned to us. "If you guys want to ride along on the chase, jump in the back. We're heading out."

You bet we did. We scrambled into the pickup bed before he could change his mind. That was the beginning of our volunteer chase crew adventure that would last nine days. Had we applied through the system, we would have been rejected as being too old. Most volunteers were young men in their late teens or twenties.

Dan, our pilot, was grouchy. When anything went wrong, he blamed the crew. One by one they deserted him, until on the last day, all that was left was Benny, his girlfriend, Joe and me.

We didn't quit because this was probably our only chance to participate in the thrill of the chase and we promised we'd help for the whole event. Being on a chase crew was hard and heavy work, but the reward was to ride in the basket during a contest.

Benny said, "The scary part of ballooning is to be up there and look down to see your chase crew stuck in a ditch on some lonely road. (This was before cell phones.) You not only worry if they can get the truck out, but if they lose sight of you, they probably lose radio contact, too. When you run out of propane, you're going to land somewhere, and they may not be able to find you."

A pilot has little control over where his balloon goes. He can direct the up and down movement by lowering or raising air temperature inside the balloon itself, but he cannot control side to side movement except by picking up wind currents.

Neither pilot nor chase crew know where the balloon will land; the pilot is totally dependent on his chase crew to find him and figure out how to bring both him and his balloon back from wherever "somewhere" is. During our nine days of chasing, "somewhere" was in parking lots, housing tracts, school yards, on a freeway, dirt roads, and in a fenced-in gravel pit.

Chase crews try to keep their balloon in sight but are hampered because they travel on roads while the balloon drifts over fields of growing crops. Sometimes we were in close pursuit when the balloon suddenly encountered a new wind current and went sailing off in a different direction. Trying to catch one particular balloon amongst dozens of similar ones makes a chase more thrilling.

Joe had his ride during the middle of the week on a basically routine trip. He was supposed to get another turn but was cheated out of it because he was needed to drive the crew pickup. When my turn came, it was *not* a routine trip.

Dan and I were in the gondola when Joe said, "There she goes," referring to the "roadrunner" leaving its traditional bomb trail.

"Let's go," Dan said, and the crew, who had been leaning on the gondola to keep it earthbound, stepped back. A long roar of the burner gave a final warning blast that sent us skimming over the ground. I looked back and saw the ground was leaving us—or so it seemed.

I felt no sensation of rising, yet the people were getting smaller until they disappeared beneath me. Even when we were flying, the only sensation of movement was a gentle swaying as if we were floating on a calm lake.

Beneath me, the earth was spread like a map. Little ribbon roads, lined with miniature houses and trees, were dotted with toy-size autos and trucks. I could see the movement of ant-size people from my former world, but the sounds of earth didn't reach us. In the sky around me, colorful jewels drifted slowly; sometimes one came close enough to wave to the passengers in its gondola. The unbelievable quiet was broken only by the irregular intrusion of short blasts of the burner—like the unexpected sharp cough of an invisible stranger.

The radio sputtered, shocking me to reality. "Dan, we're bogged down in traffic. It's going to take a while to get out of this mess."

"Shit!" Dan answered. "Hell, we've drifted off course anyway. We're out of this competition. Keep your damn eyes on my balloon."

I had forgotten about the contest, which was the purpose of our flight. I had the feeling of complete freedom. The problems of the earth had drifted away. After an undeterminable time, I saw the world I left was coming closer again. We dropped even lower until we were skimming trees over a housing tract.

"Come on down" a man called. "Breakfast is on."

"Thanks," Dan said. "We'll take a rain check."

How strange to carry on a conversation with people who were earthbound. We passed over dogs that raced around their yards, barking furiously at our intrusion. Then we drifted across a school yard where children waved miniature hands at us.

Dan broke the magic. "Chase! Where the hell are you?" he shouted into the microphone. There was no answer. "Damn! I knew they'd do this. We're out of radio range. If they have sense enough to use their binoculars, they probably have visual contact."

Probably! Wasn't he sure?

Sensing my anxiety, he said, "Don't worry. It doesn't hurt if the crew loses temporary contact. They'll find us. The problem is we're running out of fuel. I've got to land in that park." He pointed to a patch of green in the direction of our drift. "The gondola will take the brunt of the impact."

I wondered what a little gray-headed lady was doing dangling in an oversized laundry basket from a bag of hot air. Spectators were taking pictures.

Dan didn't want to open the ripping panel in normal landing procedure, so he tossed out a drag line and yelled, "Someone grab that rope!"

The spectators gawked at us. No one moved. Then we were past them and drifting over another group that had gathered. Dan threw the second, and last, drag line. "Damn it!" He shouted. "Somebody grab that line!"

A man leaped forward and caught the line. We bounced onto the ground and then shot up in the air again. We continued to bounce up and down like an out-of-control yo-yo until other spectators rushed forward to help hold the gondola down.

Dan said, "Jump out, take the line and tie it around one of the trees. Hurry up! I'm almost out of fuel. I've gotta go back up."

I'm five feet with my shoes on and the basket rim was up to my chest. Little old ladies can't jump like that. Probably no one can. The only way I could get out was lean over and fall out. A man seeing what I was doing grabbed for me; it helped break my fall, although I still landed on the ground. The same man helped me secure the tether line to a tree. (I doubt I could have done it by myself.) Dan was oblivious of my problems. His concern was getting above the tree tops where he should be visible to the crew.

When they finally arrived, Joe was surprised to see me sitting on the ground answering people's questions while Dan and the balloon were still in the air. Dan bawled the crew out for losing contact.

The chase crew loads balloon "envelope" for transport.

Dan never helped the crew pack up until we were so short-handed he had to. It was his balloon, but packing up was the crews' job. I wondered how long it had been since he was on a chase crew. We deflated the balloon, folded it and stuffed it into its protective canvas bag. Compacted into a four-foot bundle, it weighs 210 pounds, so it takes manpower to lift it into the gondola and then into the back of the pickup.

I had told Dan I was writing a story for a magazine, so he brought out the champagne. At the end of a person's first flight there is a traditional ceremony dating back to a time when pilots often landed in farmer's fields, demolishing some of their crops. Pilots carried champagne to appease an unhappy farmer. Today was my maiden voyage so they followed tradition.

I knelt on the ground where our landing was made and the crew gathered around. Dan poured champagne over my head while the crew recited the Balloonist Prayer.

Now when I see a child playing with a balloon, I smile, remembering what it was like to be completely free, drifting cloud-like through a boundless sky.

I regret that I cannot relate all the wonderful adventures we had during the 15 years we traveled while Joe worked on 52 different jobs in 27 different states. But I will add what happened in Las Vegas. We often stopped in Las Vegas to eat at their then-famous all-you-can-eat buffets. Las Vegas is where the idea of a buffet started as a way to entice people to use their slot machines.

This was a new place to us. After paying at the door, the hostess took us into a huge room with a beautiful mural on the wall. I still remember it. There was an oak tree with green leaves and colorful flowers underneath. There were three booths in front of it and she sat us in the middle booth. Then she told us to get plates at the buffet, and when we got back with our food the waitress would pick up our tickets and bring our drinks.

The buffet was spread through several rows. There was food in front of me and behind me and on both sides. Somewhere in the maze, I lost Joe. After I filled my plate, I went to the middle booth and sat down to wait. But Joe didn't come.

Then the waitress came and asked for my ticket. I said, "My husband isn't back with his food, and he has the tickets in his pocket."

"If you know what your husband wants, I'll get your drink orders. I'll pick up the tickets when I come back," she said.

I ordered our drinks. Joe never came. The waitress arrived with our drinks. "Where's your husband?"

"I don't know. He never came back." I guess I looked as worried as I was.

"Leave your food here," she said as she helped me stand up. "We'll find your husband."

Taking me by the arm, she led me around a corner and into another room with the exact same mural and three booths in front of it. Joe was at the middle booth shoveling the food in so fast he didn't even know I wasn't there.

The waitress said, "Is that your husband?"

"Yes." I felt relieved.

Still holding me by the elbow, as if I couldn't walk by myself, she led me to his table and helped me sit down. Then she patted me on the shoulder like you'd do a child. "Stay right here with your husband, and I'll bring you your food."

When she left I asked Joe, "Do you think—she thinks I'm senile?"

Joe put his "shovel" down, looked me right in the eye and said, "Isn't that better than having her think you're stupid?"

It helps to have a good sense of humor when you have a husband that goes to the wrong table.

CHAPTER **16**

Australia Travels (1990; 1992; 1999)

THIS CHAPTER STARTS with our first trip to Australia but includes a total of three Australian trips. The first was for two weeks with Lynn and our friend, Ginny. *(Lynn was divorced after the two girls were out of high school.)* We went by car and motel up the east coast after two weeks in New Zealand. The three-month trip to Australia was by ourselves from December 26, 1991, till March 26, 1992, in a camper we purchased in Australia. The final trip was a three-week trip in 1999 with Cathie, Bud, and our grandson, Travis, traveling together in one rented camper.

The first time we flew to Australia, we were met by an Australian travel writer with whom Lynn had made prior contact. She picked us up at the airport and took us to her home, where we had our own bedrooms and use of a guest bathroom.

She was a very impressive lady (about my age) who was born in England. After a bitter divorce, she brought her children to Australia to make her own way. She towered over me and issued commands that made us think of a drill sergeant. In fact, when we were not with her, we referred to her as "the sergeant."

Joe and I made a poor impression from the first. I had severe motion sickness on the drive from the airport to her house, and by the time we got there, I had to go to bed. This left her thinking I was a pampered rich woman.

Joe is Joe regardless of how important the other person is. He likes to tell jokes, some of which were inappropriate by "the sergeant's" English standards. When he told a joke about women libbers, it struck a nerve and, before Joe could say it was a joke, she was on a soapbox defending women worldwide.

In truth, Joe was more a women's libber than the rest of us, but seeing her reaction, Lynn used his salesman charm to calm her down. From then on, he was the "hero" and we were tag-alongs to be put up with. She obtained free tickets to several tourist attractions and advised us about tourist traps that saved us money.

During the three days we spent with her, she had a full schedule laid out with the exact amount of time we could spend at each. She led and we marched behind trying to take it all in along with her history lesson. I wanted to ask questions but didn't dare interrupt her preprogrammed talk or regimented schedule. She took us places we would not have seen without her. Except for the few meals we ate out (paying for hers, of course) she cooked food that was either under- or over-cooked, depending on the story she was telling.

The zoo was the last place she took us. It covered a lot of ground with pathways heading in all directions. The guys rebelled. They were determined to take their time. It wasn't long before she had left them far behind.

Trying to prevent bad feelings this last day, Ginny and I devised a plan. Ginny waited at intersections where she could see the guys, and I followed "the sergeant" close enough to see which way she turned at the intersection. I signaled Ginny which way to go, and she'd make the guys skip stuff so we'd not lose anyone. I was so busy watching "the sergeant" and signaling Ginny, I have no idea what animals were there. I don't think "the sergeant" ever looked back to see if we were still there. Maybe she was angry because we messed up her timetable.

We thanked her for her gracious hospitality, gave her a gift we had purchased earlier, and the next morning we were in our own van on the highway.

It's strange, the twists and turns life takes. Until he died, Joe stayed in touch with "the sergeant." They discovered they shared political and religious beliefs as well as a love of history. But a few years after our visit, Lynn tried to set up an expansion of his business in Australia with her as a partner. Something went sour. After that, they had no use for each and stopped all communication.

English immigrants with money went to New Zealand while the prisoners were shipped to Australia when both countries were being colonized. This accounts for the opposite attitudes of the people who now live in these two former English colonies. New Zealanders are hard-working, no-nonsense people who show up for work on time. They wanted better things and were willing to work as long as necessary to get them. In many ways they reminded me of my father.

Australians are more relaxed, especially about work. If someone says, "It's a great day for a barbie," the whole office is likely to close down for the rest of the day. (A barbie is an Australian barbecue.) Aussies seem able to consume a huge amount of beer without appearing drunk and enjoy socializing with "mates" (friends). When Lynn and Joe decided to stop at one of the famous pubs, they were welcomed warmly by the locals. Bars are a good place to make friends in foreign countries.

"The sergeant" had made out a tour schedule, but we couldn't keep up with it. We wanted to spend more time at some places, so by the time we reached Sydney, we were already a few days behind. We found a boarding house outside the city, a block from their fantastic rail system. We stayed several days while exploring Sydney.

One place where Joe and I spent more than the allotted time was the old prison museum full of history. There were only a few tourists, so when the caretaker saw our interest, he came with us, filling in details of the horrible life early prisoners had. Some were true criminals, but many were starving, homeless people who stole a loaf of bread.

From January 1788 until 1867, the convicts sent to Australia were mostly Irish. Many of them—160 on one ship alone—who were dying or dead were simply dumped overboard. Those who survived found a rocky surface crops wouldn't grow on. Seven of their eight cattle escaped.

The guards, who probably resented coming to this god-forsaken place, were free to administer whatever punishment they wished. I was appalled at how inhumanely people (in this case, the guards) treat the powerless when they have absolute control. The strong-headed Irishmen who refused to accept unjust punishment were executed.

Survival meant complete submission. Those who accepted the unfairness and served out their sentence became Australia's drovers, shepherds, farmers, beer brewers, and later miners chasing gold, silver, and opals. Australians today are as proud of their convict ancestors as Englishmen are of their lords and ladies.

Joe and I were running out of time. Our flight home involved a two-day stop in Fiji, so our friends had two more days in Australia than we did. Our plane would leave from Cairns, and it was a place we really wanted to see. We parted from Lynn and Ginny in Sydney and flew to Cairns while they drove the coastal route. This gave us time to visit the butterfly sanctuary and take the day trip to the Great Barrier Reef.

Lynn and Ginny arrived the evening before our flight home, so we had dinner together and learned what we missed by not driving to Cairns. We decided we would return when we had more time to explore this fascinating country.

We traveled for three months in a 1971 Toyota Hi-Ace camper.

That time came December 26, 1991. Joe and I flew from Singapore to Adelaide, where a couple we had never met was storing our recently purchased 1971 Toyota Hi-Ace camping van. The van had limited amenities but had an upgraded engine and a refrigerator that didn't work. We purchased it for $2,500 from an ad in *Escapees* magazine.

We removed the reefer, making room for a porta-potty. We added an ice cooler, new mosquito screening, linens, and other supplies. It needed some mechanical repairs, but we saved the $7,000 it would cost to rent a rig for 90 days. We sold it to another Escapee for $2,000 and later heard that couple added new tires and sold it to yet another Escapees couple.

Between owners, Denis and Patsy stored it on their property in Adelaide. They picked up each new owner at the airport and helped them get started. The Toyota wasn't as sophisticated as rental vehicles but was fine for boondocking. We planned to do a lot of that. Our van had only three 12-volt lights, so campgrounds offered us nothing but showers, which we could get at roadhouses (truck stops). The propane bottle with its detachable cooktop was sufficient for the limited cooking we did.

We had a five-gallon water tank with one pump-type cold water spigot but suspected it was unsanitary, so we used that water only for washing up.

With only a five-gallon tank, there wasn't a lot of grey water to dispose of. We bought drinking water in gallon bottles when we started and refilled them at homes of people we visited and at roadhouses if the water was good.

The porta-potty is designed for dumping into any toilet at caravan parks, restaurants, or roadhouses. The roadhouses (similar to American truck service stations) had free showers if you had your own soap and towel. We stayed overnight in several roadhouse parking lots but preferred "bush camping" (boondocking) because of the privacy.

Many people warned us bush camping wasn't safe. When we asked them how often they had done it, the answer was always, "We've never *done* it. It isn't safe here like it is in America." If you want to find out if something can be done, ask the man who has done it—not the one who hasn't.

Sometimes it was so hot we had to leave the van's back door open at night. This meant mosquito netting was our only protection if someone wanted to rob or terrorize us. Nobody bothered us. At night we found crossover roads that separated the right-of-way from a farmer's fence road and parked behind trees or bushes. When the heat was unbearable, we stayed in a motel with air conditioning, a shower, and a *big* bed. The bed in our van was exactly 40 ½ inches wide which meant we should each get 20 ¼ inches, but Joe took 23 inches so all I had was 17 ¼ inches. I had always heard, "camping brings you closer."

Because Denis and Patsy lived in Adelaide, we flew there instead of a more popular airport. They picked us up, having to wait at customs for a long time because I was so airsick the plane attendants took me by wheelchair off the plane after everyone else had gone. They must have wondered what they had got themselves into when they saw this obviously sick white-haired woman being pushed in a wheelchair. Yet they could not have been nicer. Patsy, a former nurse, bundled me off to bed with a cold washrag on my head, and I fell asleep immediately.

The next day I was fine, and we became fast friends with our hosts, who insisted we remain in their house while we completed the license paperwork and outfitted the van. At the end of three months of traveling, we brought the van back to be stored rent-free on their property until the next Escapees couple picked it up. All the Australians we met—and there were many—went out of their way to be helpful.

New Year's was fast approaching and we wanted to spend it on Kangaroo Island. The ferry leaves from the South Coast, east of Adelaide, and allows you to take your camper van. We would not have taken our van had we known the island roads were all "washboard" roads. Yet, by taking our van we could park in the Kangaroo Island National Park surrounded by kangaroos, koalas, and emus.

The kangaroos were so tame some would almost climb into your lap to get a morsel of the special food, which we had to purchase to feed them. What a unique way to spend New Year's Eve. No wild parties, drinking, fights, or fireworks. Simply a peaceful beginning to 1992—the year another of my dreams would come true.

New Year's Eve at Kangaroo Island campground.

On Kangaroo Island, the animals no longer fear humans; they know we have food and shiny objects (jewelry) that fascinate emus. Like kangaroos, emus will eat from your hand here, but those on national preserves run away. When protecting their young, they can be vicious.

(On our last trip to Australia seven years later, at a nature preserve Joe was wearing a bright red club shirt that attracted one emu. He stayed on Joe's heels and peeked over his shoulder every now and then to look at Joe's face. Everybody was laughing except Joe, who wasn't sure if that huge bird was going to take his shirt or if he had fallen in love and simply wanted Joe to stand still. Joe escaped.)

After returning to Adelaide, we headed north to Coober Pedy, an opal-mining town where people live underground in mines they dug while looking for opals. We drove many miles across the Outback desert where there was little to see and no traffic except an occasional road train (several truck trailers hitched together pulled by a big-engine truck). We saw very few kangaroos and only one flock of emus running away. The Outback has been glamorized, but actually it is a harsh, forbidding land that only kangaroos, emus, and sheep enjoy. In fact, if they had a choice of living where the grass is green and plentiful, even they would leave the Outback.

The summer is long with temperatures over 120 degrees; winter is cold, often freezing, with days of relentless wind. Sometimes you can go miles without seeing anything but flatland and an occasional bush or wind-twisted tree.

My fascination for Coober Pedy was that the original city was built entirely underground. We spent the night in one of the underground motels. Like other underground homes, our room had an air shaft to expel heat and bring in fresh air. Even in the middle of summer, only a fan is needed to keep the air moving and the temperature comfortable.

Dugout homes date back to miners who found they could mine year-round if they stayed underground. They began putting beds, a table and chairs, and then other living comforts in the portion of the mine already excavated. As digging went deeper, they had room to move their families into the mine, too. They paid a fee to attach to the town plumbing for fresh water; electricity was provided by their own diesel-powered generator. They dug a side shaft to dispose of sewerage.

Over time, others besides miners excavated underground homes. Entrepreneurs put in shops and motels. People pooled their money and built an underground church. Eventually, another town was built topside, but many people continue to prefer their dugout homes.

People united and built an underground church.

The main industry in Cobber Pedy is attracting tourists who want to try their hand at noodling. That means picking through mounds of waste rock (tailings) above a mine. The mounds grow bigger and more numerous as the miners carve deeper under the earth, bringing their tailings up in buckets to throw on the ground above their mining shaft.

Noodling was fun but not productive for us. We panned for gold in many old mining towns. When you pay to mine for gold, the entrepreneur "seeds" it with a few slivers of gold to inspire tourists. This wasn't needed in opal tailings because valuable opals ended up in the tailing mounds.

In addition to tourists, there were men who were seriously looking for opals and dreaming of finding an overlooked nugget that would make him rich forever. I talked with a man who came as a tourist four years before and caught miner's fever. He claims he finds enough small opals to pay for a room and food while he continues to look for that "big" one. The idea of striking it rich in gold, opals, diamonds, or the lottery completely absorbs some people.

Our original plan had been to go north from here to Alice Springs and Ayers Rock. In talking to people who had been there, Joe decided not to go for two reasons. The climb up Ayers Rock was more strenuous than we were led to believe, and I was having knee trouble. He was afraid I would hurt myself. The other reason was that flies were so thick you couldn't talk without getting them in your mouth. He had an abhorrence of flies.

So we headed back south to the coastline and then east. We stopped overnight in Melbourne to visit with "the sergeant." We planned to sleep in our camper, but since we had to park it on the street, she insisted we stay in her guestroom with the luxury of a full-size bed, a hot shower, and a chance to fill our water bottles with good water. We insisted on taking her out to dinner, and she gave us more travel advice.

Southern Shipwreck Coast of Australia.

She recommended we follow the Great Ocean Highway that circles the Shipwreck Coast. It earned its name because many ships wrecked on the dangerous reefs along this rugged coastline. Now with modern navigation equipment and big ships, there are almost no wrecks. The southern coastline route has fantastic views of crashing waves trying to cut down mighty rocks.

We stopped in Robe to see the place where Chinese workers for the Balarat gold fields were let off the ships in former days. Then they walked many miles, in single file, to avoid their boss paying the stiff entrance tariff. It was another example of the cruelty of the powerful over the powerless. Chinese men were not allowed to bring wives or children because the Australians didn't want them settling down. They wanted them to work in the mines and then go back home.

After touring Southwest Victoria, including a rainforest and Phillip Island where fairy penguins come ashore at sunset, our next stop was in Sydney. On our *first* trip there, we met Gloria and Ron while shopping for groceries.

They were so interested in America that we exchanged addresses. After we got home, Gloria and I became frequent correspondents.

When she heard we were coming through Sydney, Gloria made me promise to stay at their home for the three days we would be in Sydney. Their home was a block from a train line to the city. We intended to park in her yard and sleep in our camper, but when she said she purchased a new set of bed linens and blankets just for us, we agreed to sleep in the house. Use of a full-size bed was inviting. However, Gloria, Ron, and his sister who lived with them, were constant smokers. The only time I could breathe normally was when we went to Sydney.

Gloria insisted we have breakfast with them every morning before we went into Sydney, and she waited dinner until we got home. Strangers opening their homes to us allowed us to compare lifestyles. We learned during our travels that human beings everywhere are more alike than they are different.

Food was a main comparison. When we told them about Texas biscuits and gravy, they gagged. Biscuits to them are cookies and what we call baking powder biscuits are scones that are eaten at afternoon tea with whipped cream and jam. They introduced us to two delicious breakfasts: spaghetti on toast and baked beans on toast. Gloria's cooking was much like New England cooking: meat and vegetables cooked together in one pot with very little seasoning.

We asked about their social health care system. Both they and "the sergeant" believed it was excellent. Everyone paid into it (like our social security) and everyone could use the clinic whenever needed. People with money (like "the sergeant") also bought private insurance because it gave more immediate surgery and you could select your own doctor. At the clinic, you took whatever doctor was on duty. Not much different from our present insurance system because those *without* insurance go to the emergency room where the hospital and doctors must treat them and then try to get money if they can. Those of us with insurance end up one way or another paying for those who don't.

We also learned voting at an election was *compulsory*. Failure to vote came with a heavy fine, so protesters of forced voting took their revenge by writing down Mickey Mouse. It is usually counter-productive to force people to do something they don't want to do.

On the last day, Ron was off, so they drove us to a wildlife preserve where we could feed the animals, and their smoking didn't bother me. Their favorite entertainment was to go to a gambling club. After dinner we went there, but Australian gambling clubs are not like Las Vegas casinos.

It was serious business. There was almost no talking, no clattering of coins when someone won, no shouting at hitting a jackpot (or maybe nobody did). The computer kept track of your money, so playing the machine means pushing a button. Gloria said some clubs have card and dice games, but their favorite club had only automatic machines, keno, and betting on TV horse racing. Some nights it had bingo. I'm not into gambling, but I missed the Las Vegas glamour and excitement shared with those at the next machine.

Gloria and I stayed in touch for several years. Ron died of lung cancer about two years after we left, and then one day my letter came back with the word "deceased" scrawled on the envelope. I had not known Gloria was sick and have no idea what happened to Ron's sister. All three were much younger than Joe and me, but heavy smoking, plus living with secondhand smoke, is deadly.

From Sydney, we continued up the Gold Coast to Port Macquarie, about 250 miles north of Sydney, where there was a hospital for injured and orphaned koalas. I learned koalas are not bears, even though they may look like it to us. They are marsupials whose young are called joeys regardless of sex. Marsupials, which include kangaroos, are born undeveloped. The kangaroo has its pouch in the front because it is a ground animal that stands and jumps while koalas have their pouch on their back because they are tree animals that use their arms and legs to climb from limb to limb. Koalas are grey, not brown, although when they are young or their fur is wet, it has a brownish cast.

When a joey is born, it is about the size of the first knuckle of your little finger. This tiny baby must find its own way from the birth canal up the kangaroo's stomach or the koala's back till it reaches the pouch. It is a long, tedious, and dangerous journey. If it falls off, the mother does not know how to help it get up. Life isn't easy in the animal world.

If the koala joey makes it safely to mother's pouch, it climbs inside, where it remains secure until it is six months old. Then it is big enough to come out of the pouch and ride on mother's back. That is when it starts eating eucalyptus leaves that will be its entire diet all its life. It gets the liquid it needs from those same leaves.

There are over 2000 kinds of eucalyptus, and not all are edible. At one time, people thought koalas from one area could not be moved to a different grove of eucalyptus because they would eat only the kind they were born in. In research centers, koalas eat a variety of eucalyptus leaves and are healthier and live longer than those who eat only one kind.

Koalas get sleepy after they spend energy eating. Contrary to common belief, eucalyptus leaves don't drug them. Their sleepiness is because of very low metabolism. It also means they eat and sleep around the clock.

At two years old, it is time to find a mate. That is not difficult because once the male mates, he loses interest in the mother and baby and is ready to mate with another willing female. Koalas make almost no noise except when they are mating, and then they sound like alley cats. The joey is born a month after mating takes place.

When mother and baby finish the leaves in one tree, they climb down and go to another of their favorite trees. Sometimes this means crossing the road. Cars and trucks drive very fast on country roads where traffic is not heavy. This caused the koala population to dwindle at an alarming rate. Concerned people and volunteers established the Port Macquarie Hospital and Recovery Center to keep them from becoming extinct. It is here that a special story begins.

A new mother finished the best leaves on the eucalyptus she was living in and started across the road to check those trees. If she heard the car coming, she had no time to react. When it struck her, it tossed her to the side of the road where she lay in shock. Telltale marks showed that she wiggled her way to a pile of dead eucalyptus leaves but could go no further. Daylight came and cars zipped past. If people saw her, they assumed it was another dead koala.

That day passed and night came while she waited to die. She must have been very hungry and thirsty, but the new eucalyptus leaves were beyond reach. She beat the odds by surviving a second night, but by the next morning she was barely conscious. Then a volunteer on her way to work at the koala center saw her and stopped.

Rita, like other volunteers, watched the sides of the road for injured koalas. She walked cautiously because injured koalas, like injured dogs, can turn on their would-be saviors out of pain and fear. This koala was too near death to even try to protect herself. Talking in low cooing sounds, Rita picked up the body, and the koala opened her eyes. "You're alive! I must get you to the doctor straight away."

The koala did not appear frightened, although under ordinary circumstances she would have tried to escape in terror. At the hospital she lay quietly, too sick to move or because she knew instinctively they were going to help her.

"Oh, oh! She has a broken arm," said the doctor. "I don't think she'll make it. Too many bruises and she's very dehydrated." Someone was already dropping water into her mouth with an eye dropper. She had never tasted water before but swallowed it eagerly.

"Hey," the doctor said, "she has a joey in her pouch. She was out there at least 36 hours. What are the odds this joey would still be alive? I doubt the mother can survive with a broken arm. The bone will eventually heal, but will she ever be able to climb trees? (Koalas depend so much on their arms.) Still, I think we owe it to her to give her a chance."

While a worker held her, the doctor put a cast on her arm. "What shall we name her," someone asked. After lots of suggestions they agreed to call her Rhoda Rita. Rhoda was where they found her, and Rita was the volunteer who discovered her.

Rhoda Rita slept for days, waking only when someone gave her water or a special formula. When she finally began moving around, they gave her several more days to get used to the heavy cast on her arm before taking her to a eucalyptus tree. They put fresh leaves at the crook of the tree where the branches start because that is a koala's favorite place to sit.

"See if she will try to climb," the doctor said. "If she starts eating leaves, we can stop the formula."

Rhoda must have been longing for her favorite leaves. She started to climb, but with her arm in a cast, she could not bend it the normal way. When she fell, she looked at them as if to say, "Why don't you lift me up there?"

"Rhoda, you must learn to climb and get the leaves yourself," a volunteer said. "You have to really try." Rhoda fell again. "Come on, Rhoda Rita, find a way," a chorus of voices encouraged.

She did! She figured out how to use the cast as a brace and her good arm and legs to climb with. Everyone called encouragement as she inched her way up to those delicious leaves. When everybody cheered, she opened her beady eyes as far as she could to show her own pleasure.

Days turned into weeks, and Rhoda was happy living in her own tree where new leaves magically appeared every morning and night. Sometimes in the afternoon, when people were taking her picture, she climbed up and down to show them how clever she was.

Life was good until the night she decided to climb a little further up the tree to reach more tender leaves. Then she fell asleep. Unfortunately, her cast was between her body and the tree, cutting off all circulation to it. When she awoke, she didn't know why she could not move. The weight of the joey in her pouch was pushing her tight against the tree. She was helpless. She waited for volunteers to arrive. They always checked on her.

This morning some called up to her. She could not tell them she needed help. While they did their chores, her joey, moving restlessly, pushed her tighter against the tree. Someone noticed she had not moved and called her to come down. She didn't move.

They brought a ladder, and a volunteer climbing up to check saw the problem. Gently the volunteer carried her down. The doctor took the cast off to see what the damage was.

"It's too late," he said. "The circulation has been cut off too long. I'll have to amputate her arm." Rita opened her eyes wide to show her trust, and then went to sleep. When she woke, the cast was gone and so was her arm. Yet it hurt constantly.

She was back to formula again. She drank from the bottle while volunteers held her in their arms trying to comfort her. Days passed until finally her missing arm stopped hurting. Her joey was out of the pouch but stayed close beside her, moving only when she moved.

Once more the volunteers brought her to her favorite tree. Once more they encouraged her to climb the tree. Without a cast to use for a brace, she fell over and over. They would not give up. They kept pleading, "Come on, Rhoda. Come on, Rhoda Rita. You can do it." Once more she beat the odds!

More days passed and with each one she improved her climbing skill. With her joey riding on her back again, climbing in search of tender leaves, the world returned to normal. Every afternoon people came to look and take her picture. She was a celebrity! Her picture and story were used throughout Australia to help save the koala population that was becoming extinct.

Then another tragedy struck. For no reason they could determine, Rhoda's joey died. This was too much. Losing an arm was bad. Now she lost her heart.

They had been through so much together. Her joey gave her the strength to overcome each obstacle. Why go on struggling?

She couldn't eat, not when they brought her favorite leaves, not when they carried her from the trees and into the room with sick koalas, not even when her favorite volunteer held her in her arms and tried to feed her formula. Rhoda let the formula run out of her mouth. It was too much effort to swallow.

One day they carried her out to the big grove of trees. She wouldn't try to climb, even to get her favorite leaves. A volunteer carried her up a ladder and placed her back in the crook of the tree. The volunteer was crying and speaking soft words. Rhoda lay without moving all day and night. She didn't touch the leaves.

The next morning they looked at her from the ground and called to her softly, "Eat, Rhoda Rita. You must eat." She didn't. She didn't drink when they offered water. The day passed with her sleeping, waking, and sleeping again.

The sun was going down and volunteers came to the tree to say good-bye. They knew she could not survive another night without eating. So much had happened in her short life. Through it all, she kept trying, but when her joey died she went into depression.

There comes a time when you die or go on with life. That night, for some unknown reason, Rhoda Rita decided to live. The next morning, when workers arrived expecting to find a lifeless body, she was munching on leaves and looking at them with her beady eyes wide open as if to say, "Tell the people to bring their cameras. Here is another story to tell to the world."

Her courage made her famous and brought awareness of the need to protect koalas. We joined others in giving a donation to build koala walkways across some of the roads to protect other koalas from speeding cars. *(Update 2012: The koala population was growing until a couple of years ago when forest fires killed so much wildlife and left many koalas with burn damages doctors could not repair.)*

From the koala hospital, we drove to Queensland's banana and sugar cane country. The scenery is mainly sugar cane fields in various stages of growth because it is grown year-round. In Queensland, they burn the tops of the cane to drive away snakes and make it easier to harvest the cane. This causes a great deal of pollution, but when we were there, the major concern was cane beetles and cane toads.

Jim Gudrich told us about the toad menace. We had met Jim in Southeast Asia while touring the Japanese prison camp at the Kwai River Bridge. He was paying his respects to his fallen comrades. Prisoners during World War II were forced to build a railway bridge for bringing military and supplies from Burma to Thailand.

We became friends with Jim, whose wife was dead. Because he was alone, we invited him to dine with us. He was one of few truly educated aborigines and was head master of an aborigine boarding school until he retired. (He was in his seventies now.) He impressed us with his knowledge and said if we made it to Queensland to call him. We did.

Jim said, at one time England forced all aborigine children to go into special boarding schools where they could be taught the proper way to live. Sound familiar? It had a minimal success rate because white people never trusted them. Even after they were educated it was almost impossible to find any but the most menial jobs. Consequently, when many children grew up, they reverted to their happier ancestral way of living off the land and going "walk-about" whenever they felt the urge.

We learned to recognize them by looking at their faces. They had square powerful faces dominated by broad noses and dark eyes that looked straight through you. They never spoke first, but were quick to respond when we said g 'day.

Now, sitting in Jim's comfortable living room, we listened to the story of cane toads that was just starting to become a serious problem. It began in 1935, when cane beetles were destroying sugar crops and no control was known. Someone brought a *dozen* bufo marinus toads from Hawaii to eat the cane beetles.

However, cane beetles fly and stay at the top of the sugar cane while toads live on the ground. They seldom came in contact. This new breed of toads thrived and kept multiplying in Australia. Poison glands in the back of the toad's head killed predators. The toads also killed domestic dogs by poisoning the dog's water dish. And they carry diseases. At the time we were there, cane toads were far more of a problem than the beetle.

Since then, they have spread from Queensland cane fields in the east across the north to the Northwest Territory. A few years ago, there were over 200 million of them. Community groups formed to kill them, but their attempts have had little impact. How can you control them when a single female lays up to 30,000 eggs in a *single* clutch?

From there, we went to Tweed Heads to visit Dick and Elaine, a couple we met when they visited our Rainbow's End campground. This was the first Mormon family we stayed with long enough to learn their traditions. They had a huge store room of food, which was systematically rotated, plus a huge water storage tank, stored gasoline for their car, and, of course, a vegetable garden.

The next day they took us for a bush drive in the mountains. Elaine brought a picnic lunch. It was especially wonderful for Joe, who loved trees. Dick knew the names and history of each kind.

Dick was a trained auto mechanic who switched to being an undertaker. We stayed an extra day so he could check our camper engine that was overheating. It also needed an oil change, which was a major project with this Toyota van. Dick worked on the car most of the day while Elaine prepared a wonderful last supper for us. We spent three nights in their back yard and stayed in close touch with them for several years until Elaine died of cancer.

We headed back to Adelaide with plenty of time left, so we took the inland route. We had read about the Waterways Wilderness Park, and it sounded interesting, so we stopped there only to find it had been recently closed down by the government. However, Jean and her husband still had some tame animals running on their fenced-in land. Jean saw us looking at the padlock on the gate and came over to tell us they were closed.

After talking for a while through the fence, we convinced her it should be okay to let us look around as friends if she didn't charge an entrance fee. She gave us a guided tour, starting with the koala compound. As soon as six-year-old Buttons saw Jean approaching, she came running toward us with her arms up for Jean to pick her up.

Most places no longer let you hold a koala while you have your picture taken for a fee. Jean allowed me to carry Buttons, who put her little arms around my neck and kept looking at me with beady black eyes as we toured the park. She had been born here and enjoyed being held as much as I enjoyed holding her. When we left, Joe slipped a few dollars into Jean's pocket. She pretended not to notice.

Two other places we saw on this trip were Broken Hill, an old silver-mining town with a lot of interesting history. We had a chance to visit with a miner's wife. In addition to the danger of a mine collapsing and the future threat of silicosis (black lung disease), the labor was back-breaking and the pay small.

As machines were introduced, the work became easier, but the jobs became fewer.

Miners are very close to their mates in their work party because they depend on each other and are paid according to what the entire group produces that day. Working in underground mines means bringing your tucker (lunch) wrapped in newspaper. You carry it on you or the rats will get it. Sometimes a rat crawls up your pants looking for food. Miners never kill rats because they can detect an impending collapse. When the rats disappear, it means some portion of the mine is about to collapse.

When horses were introduced to work in mines, they were much better treated than the men; if a horse was injured or died, they had to train a new one. There were plenty of men standing in line to replace an injured miner.

Broken Hill claims to be the first Australian mine that was backfilled when it closed down. According to our newest friend, it was the only old mining town that never had houses and entire streets collapsing when the land sinks.

In one mining town, a lady, hanging clothes in her back yard, heard her baby cry and went to investigate. When she returned to the back yard, she found the clothes and line had disappeared into a huge hole where she had been standing. I don't know if that is a legend or the truth, but there is no doubt a miner and his family had a difficult life. Women lived in constant fear of a rap on the door to tell them their man was dead.

We talked to a former station manager. Instead of living on their property, many station owners hire managers with a family to live there and run the station for them. Managers in turn hire extra help when it is time to shear the sheep.

A station manager's wife is needed to be mother, housekeeper, cook, gardener, and school teacher year-round. At sheep shearing time, she helps with that as well as cooking meals for the hired hands. In her *spare time*, she makes clothes for her family. Even though she stays busy, it is a lonely life.

Americans romanticized the life of our early cowboys; Australians made living in the Outback seem romantic, too. Neither was true. Children in the Outback stayed home and were, hopefully, taught to read, write, and do simple math by their mothers.

After the School of the Air became possible (through Reverend Flynn), their education increased a hundredfold. Still, at age 13, children who want further education must go to a city boarding school. That change in lifestyle is often shocking to young teen-agers.

A father of three said, "Each child is different. Our oldest boy was ready to quit until he became involved in sports. His sister loved boarding school because she had girl friends for the first time and a social life. But our youngest boy hated city life. After a few months, he came home to the freedom of the Outback where he goes 'walk-about' whenever he wishes. He takes his gun and his motorbike and is gone sometimes for days at a time. It's the only life he wants."

Like many American cowboys who went from chasing cows on the open range to truck driving, many station owners' sons become road train drivers hauling two, three, and sometimes four huge trailers hooked together. They hauled these trailers full of goods across unpaved and unmarked Outback roads traveling so fast that they kill many kangaroos. Former station managers keep records to show friends because the kangaroo competes with sheep for what little grass there is.

Before returning to Adelaide in preparation to go home, we took a tour of Flinders Rangers National Park (500 square miles) on a "Bushman Picnic" tour. Brian's knowledge of trees, plants, and animals was amazing; the scenery was unbelievable, and the Bushman's picnic was a delicious meal he cooked over a campfire.

The Australians we spent time with were from different walks of life. There was a radio broadcaster (Denis), operating room nurse (Patsy), magazine travel writer ("the sergeant"), computer expert (John), school teacher (Ruth), aide for handicapped children (Gloria), government school administrator (Jim), machinist (Ron), mortician (Dick), wildlife conservationist (Jean), tour guide (Brian), and miners, ranchers, housewives, and Crocodile Harry who drank so much his wife insisted he dig a second dugout house just for her while he stayed in the original underground dugout.

We returned to Adelaide and spent four more days with Denis and Patsy while we prepared our camper for storage. We had time to fit in a few local museum visits. One of them was the Destitute Women's and Children's Prison Museum. We could have spent an entire day there reading the posted information. Most stories were about orphans and women left destitute because of the husband's imprisonment or inability to provide for his family. In this prison, (referred to at one time as a "workhouse"), mothers and their children were separated. They could only see each other for two hours once a month. I believe brothers were separated from their sisters, too.

We spent most of one day at the Flying Doctor Museum. Since it played such a big part in my life, I will save it for the next chapter. The last place we visited in Adelaide was a shopping mall that was nearly deserted. I don't know why. Inside the biggest building was a full-size roller coaster. Yes! In a shopping mall.

Denis and Patsy drove us to the airport, and we truly felt we were parting with family. It was our longest visit to Australia, and yet we still did not see the West Coast, North West Territory, or get to the north. When we did return seven years later, we only had enough time to see parts of the East Coast and some of the Outback in Queensland.

1999—Seven Years Later

Since the Carr family was stopping to see Hawaii and New Zealand first, the plan was to meet them at a boarding house near Sydney Harbor. We flew from Houston to Chicago where we had a three-hour wait and then to Los Angeles for a connecting flight of 14 more flight hours, arriving in Sydney 26 hours after our trip started. Neither of us had slept on the plane and all we could think of was getting our luggage through customs, into a taxi, and getting to the boarding house.

I climbed in front with the Lebanese cab driver after handing Joe the pouch that contained our passports, plane tickets home, credit card, and all of our money except for the few dollars Joe put into his wallet for cab fare. Joe got into an interesting discussion with the cab driver and set the pouch on the back seat. When we arrived, he paid the driver and tipped him three dollars. (Tipping wasn't customary in Australia at that time.) It turned out to be a wise move. The driver helped us carry the luggage to the boarding house. Both of us assumed the other had the money pouch.

After Michelle and Bill showed us our room (upstairs) we went back downstairs to sign in. That's when we discovered neither of us had the pouch with our money and documents. Bill called the police to report our loss, but there are many cab companies at the airport and we couldn't remember the name of ours. The police agreed to come "when they had time."

We were both dead tired but too worried to sleep. Joe called the United Airlines desk, but she couldn't help us. She did tell us not to worry because *usually* these things turn up at the police station sooner or later. After a while, the police arrived and started interviewing us.

While this was going on, Michelle got a phone call from the cab driver whose last passenger had found our pouch on the floor where it had slid off the seat. He said he was already on his way to bring it to us. As you can imagine, he got a very generous tip. It was noon before we got to bed.

When the kids arrived the next day, Michelle booked a one-day trip for us to the Blue Mountains. In the meantime, we explored Sydney, took a two-hour harbor cruise and went to the aquarium. Here we walked through tunnels where the water and fish are swimming on both sides and over your head giving the illusion of walking under water. Our grandson loved it.

Although the camper van we rented was bigger and newer than our old Toyota, it had many mechanical problems and was overrun with ants. But it slept five. We made the mistake of driving too far between campground bookings. Bud expected to cover as many miles a day as he can in the U.S., but on Australia's rough two-lane roads it wasn't possible.

We stopped at the koala hospital and learned Rhoda Rita had lost two more joeys before developing a severe case of "wet bottom," a common koala disease that requires euthanizing them. Her life was filled with one tragedy after another. Yet she showed resilience and courage after each setback.

We drove past cane fields that were in bloom; the feathery-grey cone-shaped blossoms looked blue-grey when the sun hit them. From a distance, it looked like frosting on a green cake. At a sugar plantation museum, Cathie found a flyer about a four-day train trip into the Queensland Outback.

People *had to* book ahead, but we decided to try anyway. The grouchy ticket seller, at the Savannah Lander train station, wanted to book us for the following week even when we said we would not be here that long. He suddenly changed his mind when I said, "What a shame. I'm writing a story about Australia for an American magazine. I wish I could have included this train trip." He gave us a discount on five tickets plus some "dignitary" gifts.

The next problem was what to do with the camper van and our luggage. I spotted a sign advertising Joe's Lock-Up for $10/night. The owner was very nice and gave us a key to his office so we could use his shower, toilet, and his phone to call a taxi to the train station the next morning.

The Savannah Lander consisted of two train cars, each one with an "engineer" (driver). Passengers could change seats when desired to talk with other people. We all became friends. We were the only Americans.

At lunch the train would stop somewhere that food was sold and wait for us all to get back on board. At night it would stop again, and we'd all get off and stay at one of the boarding houses in town. The boarding house would provide dinner and breakfast. Then we'd gather at the train for another day's trip.

In that part of the Outback, we saw beautiful birds, including eagles, hawks, and many colorful species of the parrot family. Having both driven and taken a train through the Outback, the train was more fun. All five of us could enjoy the scenery that was limited to only one passenger in a motor home.

Also, driving a camper van leaves you without assistance if you have engine trouble. After an unexpected rain, roads can wash out leaving you bogged down in mud. The train stops at various points of interest along the way. We saw special grass that rotates in a clock direction when it rains, a soap bush, and picked up a souvenir "bush clothespin"—a hard seed pod with a crack down the center.

The engineer driver on our car let Travis sit up front in one of the engineer seats. The driver's mother, Betty Walker, was also on this trip. We didn't know who she was, but because she was the only single passenger, we invited her to sit with us at meals. We do that whenever we see someone who is alone. After we returned to America, I corresponded with Betty. When the Savannah Lander won a national honor, she sent me a copy of the certificate with her son's name on it.

Another place that impressed me was a stop in Chillagoe, an old mining town in the Outback. It had a hospital on top of a hill, which was actually the only doctor's house. He had installed an operating room. Unfortunately, most people sick enough to need a hospital couldn't walk up the hill, so they had to find someone with a horse (and, hopefully, a wagon) to get them up there. According to the *legend*, one morning when the matron (nurse) got to work, she went looking for the doctor and found him in the operating room where he had just removed *his own* appendix. She arrived in time to sew him up. I love local legends even when they contradict what is believable.

Other popular stops were a marble quarry, which Joe was very interested in learning about, and a bush camp by the train track where a nice young couple served "billy tea and tucker." It was in the middle of nowhere, surrounded by brush. The tea was made in a big pot over an open campfire and dipped out with a long-handled ladle into a tin cup. They had homemade cakes and handcrafted souvenirs for sale.

We took a side bus trip to see lava tubes with a tour guide who told stories about the various plants or animals along the way. Each time he ended with, "But we didn't come here to learn about that. We're here to see the lava tubes." And he would start the bus again. In between stops, he recited neat poems he had made up himself.

When the train got back to Cairns, Bud booked a trip to the Great Barrier Reef. I had been there on our previous visit and was seasick the whole day. The water looked even rougher today, so I stayed behind. Unfortunately, the water was so cold they couldn't scuba dive or even do much snorkeling.

Then it was time to head south, turn in our camper van, and prepare to fly home.

On the way, we made one more impressive stop at Paronella Park, where many paths wander through a rain forest that was like walking in a fairyland. When Jose Paronella moved to Australia from Spain in 1913, he decided to buy 13 acres of virgin scrub along Mena Creek. In 1929 he built a castle that had a great ballroom and stonework balconies and lovely gardens. He planted seven thousand trees, excavated a tunnel through a small hill, and built small stone bridges. He opened it to the public in 1935. After he and his wife died, the children sold it to another couple who decided to keep alive Jose's dream of leaving the world the legacy of a fairyland.

That was our final tour before leaving Australia for what was to be our last trip there. We saw enough of Australia to know that it is many things, just as America is. From lush forests, coastal beach towns, and rugged southern coast to the heart of the Outback, you see weird animals, huge birds that cannot fly, trees that do not produce shade, and flowers without perfume. Those who prefer city life can find lovely little towns and large, modern cities. If you are interested in history, Australia's is fascinating! Australia was all we hoped it would be. We found many strangers who opened their homes and hearts to us. I wish we had had time to see every part of it.

However, America will always be home, and there are many places here we haven't seen yet.

CHAPTER 17

Miss Kay's Dream (1992--)

WE ALL WANT to be in control of our life—to be independent. Some of us will have a sudden heart attack, stroke, or accident, and go quickly without time to worry about what is happening to us. Others will live years with some degree of disability that takes away our independence and control. For Escapees members, this will be the time they must hang up their keys.

Building home-base parks answered the question of what members could do when one of them became ill and needed a doctor close by or became too old to drive safely. They were not always the answer if a spouse died or was physically unable to care for the sick one.

Our philosophy of caring and sharing does not include taking on the responsibility of a neighbor who needs *continual* help. The result was members sold their RV and home base and moved in with a family member whose busy life might already be at the breaking point. The alternative was going into a nursing home when they could have remained independent with just a little help.

Since we encouraged so many people to trade their house for a home on wheels, I believed it was our responsibility to help members remain independent as long as possible. If all they needed was "a little help" to do that, we should provide it.

Because people were living longer, more people were diagnosed with Alzheimer's disease that required a 24/7 caregiver. My dream was to find a way to help those members cope with their new restricted life. The trouble is, most dreams take a lot of money, determination, knowledge, and faith. I had the determination and the faith but neither the money nor the knowledge.

Then in March 1992, before we left Australia for the second time, a man I never met showed me how to do it. He was an Australian hero and Presbyterian minister named John Flynn.

After being ordained in 1911, his assignment was the Outback of Australia. Most Australians live on or near the green coastal areas. As you move further from the habitable coast, you go deeper into the bush. As you get further from any coastline, the more desolate the land becomes until you reach that two million square miles called the Outback.

Like all Australians, Flynn heard stories about hardy souls spending their lives scratching a living from the infertile desert. It filled a young man's head with dreams of freedom.

Flynn had a rude awakening now that his mission was to ride his horse for several hundred miles to get from one station to another. Now he realized how isolated these people were. As he traveled cattle track "roads" that washed out when it rained, or were obliterated by red sand when the wind blew, he began to understand how primitive their life was. Lonely graves along the way told tales of people dying because they could not get medical help. No one seemed to care, so Flynn made this his mission in life.

Think of the problems. It took days to ride from one station to the next. How could a sick patient get to one of the *two* doctors who covered the entire Outback? There were no telephones. How could you get help for a broken bone or an infected cut?

Flynn decided the only practical way was for the doctor to fly to the patient. How many doctors have a plane? In what medical school does the curriculum include flying lessons? If by some miracle the doctor had a plane and knew how to fly it, with no means of communication, how would he know where the patient was and what medicine to take with him?

Flynn was not a doctor. He did not have an airplane. He was not a pilot. He had no money. Like Korczak who dreamed of turning a granite mountain into a statue of an Indian, Flynn's dream was "impossible." Everyone said so. Like Korczak, he didn't listen. He was determined to solve the unsolvable problem.

His first goal was to get the government to provide every station with a wireless hand-cranked (ham) radio so they could communicate with each other *and with a doctor* by tapping out the Morse code. Picture it. With one hand, you turn the hand crank while you tap out the series of dots and dashes that spelled words. It was difficult and limited, but people at isolated stations could finally communicate with a doctor and with each other.

Next he convinced the government to provide each station with a medicine box containing the most commonly used medicines in numbered bottles. Using Morse code, the rancher could reach a doctor and say: "My five-year-old child was bitten by a red spider and has a high fever."

The doctor could give instructions: "Give him one pill from bottle number 12 now and one from number 15 now and every four hours until tomorrow. Call me tomorrow and tell me how he's doing."

Flynn's ham radio communications were also the beginning of the School on the Air, which still is the only means for children in the Outback to get a basic education.

Money was one of the major things Flynn needed to buy planes and hire pilots to make his program handle serious medical problems now that the simpler ones were solved. He told people what he needed and asked them to send whatever money they could. Through donations of all sizes from a vast number of people, Flynn accomplished even more than he originally hoped to do.

That was the beginning of what became the world's model Flying Doctor Service. Today, Australia has a very sophisticated system of flying either the patient or the doctor to each other. Many countries are using a modified version of airlifting a patient by helicopter with a team of experienced technicians and modern equipment to the nearest trauma hospital.

This summary of the story I read at the Flying Doctor and School of the Air Museum in Adelaide made me realize I could use his money-raising method to make my own dream a reality.

When Joe and I returned from Australia, it was time to head to the Escapade in California. At a happy hour with six invited people, I shared my plan for a place members could park and get the assistance they needed. I picked six people that I knew wanted me to succeed but honestly believed it was impossible to accomplish.

Then we went to dinner together, and by the time we headed home, all seven of us were committed to being the organizing board of directors for a separate *nonprofit* corporation with its own governing board. The seven people were Don Alexander (a retired IRS executive), Bud and Cathie Carr (head of Escapees), Norman Crook (a minister), Joe and Kay Peterson (founders of Escapees), and Carol Richards (an attorney).

Carol wanted the word "care" to be part of the name. During the night, I woke up with the thought that CARE could be an acronym standing for "Continuing Assistance for Retired Escapees." Retired now meant "retired from the road."

At closing ceremony, I presented the plan to all the attendees. I explained why we needed a safe place for members to go and get the help they needed. The only way we could afford to do it was by pooling our pennies and dollars. When the meeting ended, crowds of people tried to shove money in my hands. I could not take it until we were officially incorporated, so I asked them to wait until we asked.

Before we left for our various travels, the founding board visited *On Lok* in San Francisco where the idea of adult day care started years before the government became interested. Someone with influence heard about their program and convinced the government to look into it.

Our government begins every project with studies and surveys. A survey determined that 80 per cent of those currently in nursing homes did not need the 24-hour assistance the government was paying for. The main things the elderly need are a balanced meal prepared for them, transportation to doctors and other service providers, sometimes help with bathing and dressing, supervision for those with memory problems, and a chance to socialize with each other.

Home health organizations sprang up all around the country along with a senior center where people could socialize and have one prepared meal a day. Then most senior centers found volunteers to deliver that meal to seniors unable to get to the center.

In 1990, the government started a pilot program, called *Pace,* in several states to test the benefits of adult day care for those who didn't need 24-hour care. One of these test programs was in El Paso, Texas, on our route home.

One of the biggest advantages was for adult children trying to care for parents with dementia and work at the same time. Day-care programs made it possible for more families to keep a spouse or parent at home. Adult day care is based on the same concept as child day care. The difference is that as time goes on children need less help whereas the elderly need more.

While Joe and I spent the remainder of April and most of May looking for land in Texas on which to build our new facility, Carol set up a nonprofit corporation called Escapees CARE, Inc.

In June 1992, (just two months after my Escapade announcement) we were able to accept donations. Twenty years later, we still depend on donations to allow us to keep our rates low enough that even those on limited pensions can participate.

About the time we acknowledged land was too expensive or not in an appropriate place, Escapees purchased a large piece of property adjoining its Rainbow's End Park. Both boards got together and worked out a plan whereby Escapees would lease 10 acres of its new property to CARE for a low annual fee, which Escapees has donated back to CARE. The Care Board was determined not to build anything until they had the money in hand to pay for it.

CARE was built without any government help because the purpose of our project was to help our members rather than the people in town. However, after the building was finished, we accepted local townspeople for the licensed weekday adult day care program but not the RV resident assistance program.

CARE site for Escapees residents.

The construction began in 1993 when project manager, Leon Culbertson, and his volunteers, began leveling what had been a cow pasture into a building site. They built roads and installed utilities and a septic system. On each of the original 21 sites, we built a small storage building with a cement patio.

There was no "down payment" except that participants paid a month in advance. At the end of each month they could leave, without losing any money, if their reason for coming was solved. If a participant died, the spouse had the option of staying on or going back on the road. (This was what I had promised myself back in the 60s that I would do if I ever had a chance to help the sick elderly.)

Three years after I convinced people that it was possible, six sites were finished and two were already occupied, even though we had no building and no paid staff. Volunteers provided transportation and sat with dementia patients while the spouse went grocery shopping or attended to other needs.

In 1994, our first paid employee, with the help of a doctor in the club, put on a weekend health fair in partnership with our local hospital. This became an annual week-long event that allows park residents, travelers, and townsfolk to select the blood tests they want for a minimal cost; the results are given to them to share with their doctors. A local doctor offers his time to interpret the results each year and answer questions travelers may have.

Health fair week has free seminars given each day by local health providers plus some evening entertainments. The only part of the health fair that CARE receives donations for is one dinner with entertainment and an auction of donated baked goods. Many local businesses help sponsor it and provide door prizes.

That annual event, along with two garage sales of donated items, raffles of donated crafts, and a booth at Escapades are the only fund-raisers CARE currently puts on. Chapters and BOF (common interest) groups hold fund-raisers at their rallies; along with individual donations and money or property left in wills, they keep CARE functioning at a cost far below what other facilities charge.

Ribbon-cutting ceremony March 12, 1995 was official opening.

The building was completed and furnished in March 1995, but it is impossible to say when it began. The seed was planted when I was playing baseball behind the poorhouse. It was fertilized when I worked at nursing homes in California. It took root in 1992 when people began donating their pennies and dollars. It has continued to blossom. If you don't see the importance, ask those residents who were desperate for help when they parked their RV on a CARE site.

On March 12, 1995, we held a ribbon-cutting ceremony. A social program was established by volunteers. In May, Richard and Helen stopped their travels long enough to start a lunch program so that anyone, whether Rainbow's End resident or a CARE participant, could be sure of one nutritious meal a day for a small donation. That was the beginning of a meal program that now provides three meals a day, seven days a week—including holidays—free to residents and adult day care participants and for a nominal fee to anyone staying at Rainbow's End Park.

We soon learned people needed more than a social program and volunteers to drive them to appointments. To provide more, we applied for an Adult Day Care license. During the three years since we were incorporated, both the state and federal governments had been adding rules that were originally intended for 24-hour nursing homes.

Our building, built under the first code, now required expensive changes—such as widening the corridor by an inch and lowering sinks by two inches. It seemed as if the inspectors were looking for reasons not to give us a license. We had the chicken-egg problem too—we could not get a license without the required full-time *paid* staff, and we couldn't afford that staff until we had enough participants.

Our program was different from any other adult day care in the country because our participants live in their (RV) homes on CARE property. That meant they were our responsibility 24 hours a day, seven days a week, including holidays. All other *day care* programs are five days a week with "clients" leaving at night.

October 4, 1996: Rachel Ludwig, our first RN facility director, had the experience and knowledge needed to obtain our license to operate an adult day care program. It came with the provision we accept town residents as well as Escapees.

From the beginning, our corporation was registered as nonprofit, but the IRS decided we did not qualify for tax exemption because our primary purpose was to serve our own membership. This meant people were less likely to make big donations or include CARE in their will because there was no tax benefit. Thankfully, this did not stop the true believers from donating.

By 1998, we were running out of room, so we doubled the size of the original building, expanded the size of the required covered outside area, and built more sites. The expansion enabled us to add two larger shower rooms and more toilets. Several industry-related businesses gave donations toward this, and more chapters held special fund-raisers. Escapees chapters and BOF groups receive a patch for every $1,000 donation they raise. They are proud of their patches and display them at Escapades.

In August 2000, we were finally able to get tax exemption for donations. This was primarily because of the tenacity of Gene Lacy. Now people wrote larger checks and included CARE in their will.

Today Escapees CARE, Inc. is a nonprofit Texas-licensed tax-exempt (501)(c)(3) adult day care facility. Although residents pay to live there and get the services they need to remain independent, the amount they pay is about 64 per cent of the actual cost of their care. The remainder is subsidized by donations and fund-raisers.

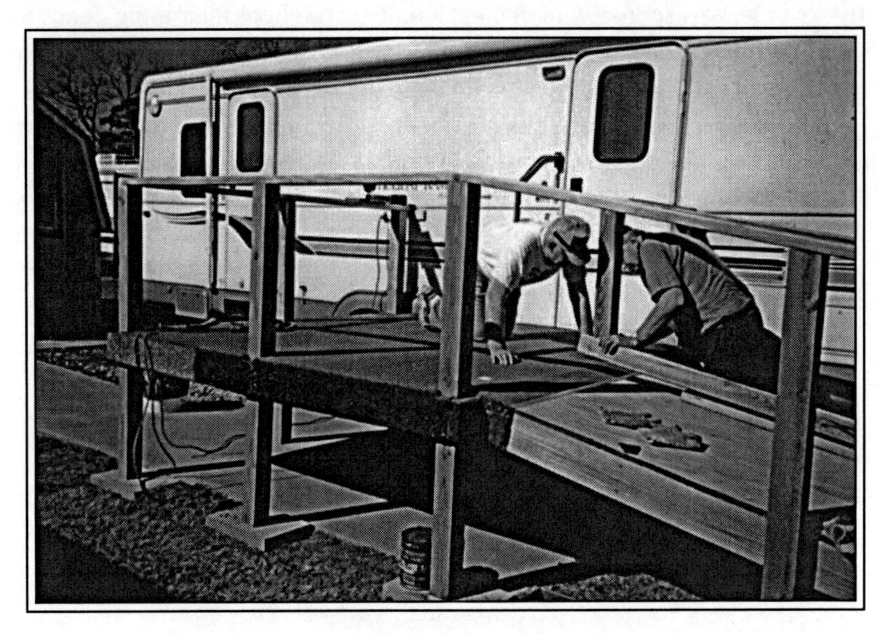

Volunteers have saved CARE money in many ways over the years.

From the beginning, volunteers have played an important role in keeping expenses affordable. They do many things, from building ramps for newcomers who need a way to get in and out of their trailers to helping cook and serve meals, drive residents to doctor and other appointments, and answer the emergency help phone after hours and on weekends.

In 2004, CARE purchased three acres adjoining the former property for future expansion. In 2005, an old house on the property was gutted, rebuilt, and made into a guest cottage that is rented to visitors or used by potential participants who want to check out the program before they have someone drive their rig to Texas. Many volunteers helped, but Don Burks and his mother, Pearl, paid for a big portion of the building material and landscaping.

There is no other adult day care facility like Escapees, CARE, Inc. Spouses can be an active part of the program; residents can keep their beloved pets and can come and go from the activity building to home whenever they wish. All the advantages of assisted living are available to the residents.

CARE's licensed adult day care program provides town and local people with the same nursing care as traditional programs because they are all under the same state and federal licensing requirements.

Many of our members have found this the haven they were looking for. They can leave a spouse with dementia and escape from their tiring duties for up to 10 hours a day on weekdays. Many would have had to go to a nursing home much earlier in their illness if it were not for our adult day care center.

Like the Escapees Club itself, the CARE program has been a total success because of the continued support from donors and volunteers.

For more information:
www.escapeescare.org

CHAPTER 18

World Travelers (1973—1991)

BY 1987 OUR daughter, Cathie, and her husband, Bud, were running the club, using us mostly as consultants. Having family members take over gave us an opportunity to travel outside the United States, sometimes in RVs and sometimes on bus tours.

Our first airplane flight was to Hawaii where we signed up for a programmed tour. The next day a company bus picked us up at our hotel, and we spent the entire day driving around Oahu. Our driver was especially good, and Joe appreciated that he knew the names of the flowers and trees. He entertained us with stories, legends, and historical facts.

Our hotel was the first drop-off when that wonderful tour ended. He announced over his loud speaker, "If you've had fun, it is customary to show your appreciation."

Since we had paid a high price for the tour, it never dawned on either of us that he was "hinting" about a tip. I took him at his word. Since we were sitting in the front seats, I jumped up, faced the other passengers, and said, "Yes. Let's show our appreciation. Let's have a hand for this fantastic bus driver." I then led them all, and we clapped long and loud.

Then I started singing, "For he's a jolly good fellow" and soon everyone joined in. Joe said it was to drown out my off-key singing, which does have some original notes. When the song ended, we all clapped again. I was not trying to be a smart-aleck, although some passengers may have thought so. It simply never crossed either my mind or Joe's that passengers give the driver an additional tip. We were naïve travelers except by RV.

We were the only ones getting off at this first stop. The driver jumped up and was standing on the curb. He graciously extended his hand to help me down. Then he extended his hand again to Joe who thought he wanted to shake hands and did so vigorously. During the hand-shaking we both told him again what a great guide he was and how much we enjoyed the tour.

I wonder what was said on the bus after we left. There is a lesson here. Sometimes we think giving a hint allows people to read our minds. Why didn't he come right out and say, "If you've had fun, I'd appreciate a tip?" We'd have given him a generous one.

On retrospect, maybe we gave him more than whatever dollars he expected. We gave him a story about two naïve Americans who thought showing your appreciation meant clapping and singing. Telling that story would make future passengers laugh, and if there were other unsophisticated travelers, they may get the hint.

We went back to Hawaii again and saw three of the islands. But when someone mentions Hawaii, it isn't the beautiful flowers or the gorgeous scenery that comes to my mind. It's that first bus tour that prepared us for future overseas travels.

Our next overseas trip, in 1990, was with our friends Lynn Rogers and Ginny McDonald. We flew first to New Zealand and rented their largest camper (mini-motor home). We drove around New Zealand's northern island first. In September, the rolling green hills and valleys were dotted with grazing sheep. We were told there are 60 million sheep in New Zealand, and since the seasons are the reverse of ours, they were having their spring lambs. They seemed to jump with the sheer joy of being alive.

We took a ferry across to the South Island, which has more mountains and forests than the North Island. Here the view started with scrawny trees and brush, then changed to bare mountains and then broke again into a tree-enveloped road and forested cliffs. Sheep and cows were running wild on parts of the South Island.

There are two things I remember most about New Zealand. One is the luge ride. We parked at the gate and took a cable car to the top of a steep hill where the luge is one of two thrill amusements.

A luge is a sled on wheels, directed with a handle bar. To slow the speed or stop it, you can push the handle all the way forward or pull it all the way back. The track was a hazardous one made from cement (with no side rails except at the very top). There were thrilling (read that dangerous) curves from the top of this high hill down to the entrance gate. They did not provide helmets and assumed you knew what you were getting into when you paid your money.

Joe tried his best to talk me out of going. Sometimes Joe knows me better than I know myself. I should have listened to him. Ginny, who was as small as I and 10 years older, had already taken off followed by Lynn. If she could do it, surely I could.

With great reluctance, Joe went over the instructions again. The part that stuck in my mind was in order to slow down, you push the handle all the way forward. Pushing it part-way forward doesn't slow you at all. What I failed to consider was that my arms are short, so when I thought I was pushing all the way forward, it wasn't connecting—I was going faster. I completely forgot I could stop by pulling the handle *back*.

I took off like a shot and Joe grabbed a luge and took off, afraid of what he would find. He found me around the first curve where old tires were piled in the middle to help you follow the curve. I hit the tires head on, flipping me over my luge, and I was sitting dazed in the middle of the tires when Joe reached me.

After a quick examination for injuries, I looked down and saw there were no sides on most of the cement road. Those tires allowed me to escape death.

Joe insisted I leave my luge on the tires and walk back to the entrance. "Stay close to the side so you won't be hit if another luge comes whizzing by," he cautioned.

The lady attendant saw me coming and screamed, "You can't walk up here!"

"Want to bet? If you want your stupid sled, it's in the pile of tires."

"I don't understand Americans. Your friend—the *older* lady went down fine."

"Yeah. She jumps out of airplanes, too."

I decided there are two kinds of people in this world. Those who live for adrenaline rushes are like soaring eagles. The rest of us are chickens who will never fly. Yet life on the ground is good also.

Besides its beauty and friendly people, the other thing I remember is New Zealand's wonderful campgrounds where everybody cooks meals in the big kitchen furnished with all necessary appliances. It is a great way to make new friends while you cook, eat together, and wash your dishes in another common area with big sinks.

On our way home, we had a two-day stopover in Fiji, so we rented a car to explore Fiji on our own rather than taking the bus tour. We had a moment of hesitation when we passed through customs and saw soldiers with guns watching everyone. We arrived at 2:30 a.m. but it took another hour to clear customs and rent a very small car. They gave us a map, which was of no value because it was not in English, and there were no street signs on any roads except in the tourist area. I don't mean English signs. I mean *no* directional signs. Most tourists take the bus tour or hire a guide.

The car was small and we had so much luggage that we had to sleep sitting up until sunrise. We started north on the coastline road, realizing it would eventually bring us all the way around the island. What fun to watch the island wake up!

Nobody had cars, so ours drew lots of attention. Curious children with big brown eyes stared at us.

Some men walked to the cane fields carrying a short-handled machete while others led one or more oxen who would graze on cane leaves when not pulling a plow.

We saw young barefooted boys leading one or two cows down to a pasture and women washing clothes in a bucket of water. We passed little villages with tin or wooden one-room shacks and straw roofs. There were chickens and goats in each village plus a cow or two.

We talked with a cane farmer whose son was working in the field with two oxen pulling a primitive plow.

The map showed a small town that looked close to the coast road, so we headed there. When we came to an intersection, we had no idea which road to take. There was a bench for people to sit on while waiting for one of the few buses (old cars) that came by at some undeterminable time. A native waiting there waved and smiled so broadly his ivory teeth, surrounded by jet black skin, reflected the sun.

We stopped, hoping he could help us. He spoke English and, with the help of the map, gave us the needed direction.

"That's where I'm going," he hinted.

Our mini-rental car was full of suitcases, but he was skinny and by putting one suitcase on my lap and one on his, we were able to make a tiny corner by one window. He chattered all the way telling us about village life.

"Each little village has its own chief who wields discipline (usually beating with a stick) and keeps the young people under control. We don't burn sugar cane here because our cattle eat the green tops. Our entertainment is singing, talking, and dancing (no electricity, televisions or radios). We don't mix with the Indians who migrated here. We never see white foreigners on this part of the island."

When we got to town, he directed us where to turn. We soon realized he was taking us down every street in town. All the while he had his head out the window waving and calling to his friends. He was having such a good time, we circled the town twice before we let him out.

He stood grinning and waving both arms until we were out of sight. He welcomed the ride, but his real enjoyment came from having his friends see him riding in a white man's car. I'll bet he still tells anyone who will listen about that day. Later, I remembered our Polaroid camera and wished I had taken a picture of him with us and the car to give to him. He would have treasured it.

In an even more desolate part of the island, we came across a tiny settlement with four houses. Two adorable little native girls stood outside watching our car as we approached. Communication was difficult as none of us spoke the other's language, but with gestures, smiles, and hand signals, we asked to see inside their primitive shelter that served as a house. The mother let us go inside.

This time I took a picture of the family and then two pictures of the little girls. I gave the mother the family picture and each of the girls their own pictures. I had performed some kind of magic, and I'll bet they still have faded photographs they show to their own children.

Sharing a stranger's life is the best part about traveling to me. Making this trip "on our own" was a magical experience. We saw a different Fiji than bus tourists see.

We went to Bangkok, where years ago my sister worked at the U.S. Embassy. This is where she met her British husband. She had told me that Thailand was beautiful. She was right.

The primary religion is Buddhism, so there were many gaudy temples decorated in gold and valued stones and each contained statues of one or more of the different Buddhas. (Both men and women were portrayed as Buddha.) It seems Buddha was more a way of life than a deity.

The floating market exists in no other place in the world. Merchants live along the banks of the canal and put their goods (primarily fruits and vegetables) in long, wide canoes and sell from the water. Buyers can rent boats and go up and down the canal trying to get bargains. Floating markets have existed since the town was established many years ago. The waterways around the floating markets are like streets in a city with side canals going in many directions.

Typical Thailand meat market.

They also had market areas along city streets where many items, including fresh meat, were sold from stalls. I recognized dogs and monkeys among mysterious cuts of meat. The vendor cuts off a piece and hands it to the customer in a sheet of newspaper unless buyers have their own tote bag.

Ducks and chickens were sold alive, but the vendor would oblige you by wringing the neck if you were queasy about carrying a live bird by its legs while you shopped. It seemed as if there were flies everywhere, but it didn't seem to bother anyone except us.

One evening we watched a woman washing plates outdoors at the back of a restaurant. She used two large dishpans. In one was greasy water that may have been hot when she started, but from the grease floating on top was cold now. After she ran her hand over each plate to be sure there were no scraps left, she put it into the second dishpan of rinse water. A young child helped her. He reached in, got a plate out of the rinse tub and dried it with a cloth he must have been using for days.

Clean plates were hustled to the kitchen inside by another young boy. Our hotel warned us not to eat food from vendors, but the signage, tables, and chairs announced this was a local restaurant. After seeing this, we ate only at hotel restaurants, missing the "taste of Thailand" food by being squeamish.

In Thailand, overcrowded cities and rural life were very different. In 1991, 80 per cent of Thai people were still farmers living in rural areas. Rice farming was a back-breaking job involving the whole family. In rural areas, crops were watered by hand, walking between the rows with a dipper and buckets of water.

The main entertainment for farmers was to go to the temples, catch fish, do handicrafts, and attend cock fights. Instead of going to school, poor children work. When grown they can save their money and go to an adult school to learn to read and write in Thai and do basic math. Even those whose jobs deal with tourists know only necessary words. Smiling and nodding, even if they don't know what they are agreeing to, is their communication with foreigners.

We saw no violence or gangs, although political turmoil was already starting. The old days of a ruling king were gone. *(Since 2006, the country has been in constant political battle for power. Off and on, it has been a democracy. Attempts are still being made to change the country's government with the result of frequent street riots. I'm glad we went when Thailand was still peaceful.)*

The adventure I remember most was meeting a man who took us and a Chinese family in his van for hire. Lee was an impressive young Thai with little formal education. He lived in an orphanage where missionaries taught him to read, write, and speak Thai. When he was 10 he ran away and joined the Vietnam army—and has scars to prove it. During eight years of fighting in a war, he learned to speak and write English from American and British soldiers and speak both Chinese and Vietnamese from other soldiers. At age 18, he spoke four languages, so it was easy to become a tour guide.

Lee was just over five feet and thin. His dream was to save enough money to buy land and start a fruit farm. After that, he planned to start a market for himself and other farmers who lived in his village. However, before he started that life career, he was preparing for a two-month trip to Alaska where he had a job waiting on a fishing boat owned by an American he met on one of his tours. He proudly showed us the passport and plane ticket he already had for this great June adventure.

I remember him most for his beautiful smile and his singing. Sometimes he sang softly to himself; other times he sang in loud melodious tones. He smiled whenever anyone looked at him. He was a perfect guide for exploring the hill country in northern Thailand.

After lunch in Chiang Rai, we boarded a long-tail boat (Thai adaptation of a canoe with a motor on the back) and headed up the river to the remote Karen Village, which provided elephant transportation through the jungle. We climbed a platform to get on the elephant's back where there was a wooden seat with a rail along the back, sides, and part of the front. Joe and I shared the rickety seat and a driver sat on the elephant's head to steer him. We rode in a procession of a dozen elephants, each following the one in front of him. It was an uncomfortable two-hour ride through the jungle to a remote tribal village.

When the elephant climbed high hills, it threw us backward against the railing giving me a sense of security. Going down hills was very different; it took all my strength to hold onto the side rail, and even with Joe's arm supporting me from in front, I had the feeling of falling.

Lee walked the entire distance. We could often hear his voice singing to the forest, to the animals, or perhaps to his God.

Our destination village was where nomadic tribes lived. Many houses were built high on stilts making a cooler place under the house for children to play while adults did crafts. At night the family's animals—dogs, chickens, water buffalo, and sometimes pigs slept there.

Adults wore pieces of mismatched clothes, but many children were naked. Lee said they considered clothes an unnecessary luxury. They lived off the jungle and traded for things they needed. Occasional tourist groups like us gave them a few dollars. "I wish I knew they needed clothes," I said. "I would have brought some."

"Mi pen ri," answered Lee. "It means, don't worry—it is not important. These people are happy because they are free. They go where they want."

A van was waiting to take us back to Chiang Rai several hours away. The elephants quickly disappeared back into the forest with their drivers. After a while, we came to a larger village where there was a public toilet. Ladies walked through the men's room to the back, which had two "squat" toilets. That means you spread your legs with your feet planted on either side of a hole in the cement, squat and hope you hit the hole instead of splashing your clothes or feet. A woman was selling one square of rough toilet paper if you had the correct change. Joe had all our money. Travelers learn to adapt to the customs of the country.

As we drove home that night, I remember seeing a young man, possibly a teen-ager who had an accident. His wagon and bicycle lay demolished beside the road with contents scattered. He sat cross-legged in the center of the destruction staring at his loss. It was close to dusk. Was he so devastated that he did not know what to do? Did he hope his family would find a way to come and get him?

I could not sleep for worrying about the man's loss of transportation and possessions. That is the part I hate about tours. Everything is on a tight schedule. Had we been traveling on our own, we would have stopped to see if we could help and at least given him a ride somewhere.

From Bangkok we flew down to Singapore to spend a few days, including Christmas Day there. We saw Christmas decorations in Hong Kong, but Singapore is not a Christian nation. I was surprised and pleased to find our hotel decorated with red poinsettias, balloons, greenery, and colored lights. The stores on Orchard Street (main shopping district) were as festive as Las Vegas.

Actually, Singapore city takes up the largest island but it also includes about 60 small islands that are under the same parliamentary republic government. It attained the rights of a British protectorate when Japan was defeated in World War II but did not gain full self-government rights until 1963. It became an independent country in 1965.

Until then the government was corrupt, drug-ridden, and controlled by the mafia. Then the people elected a new president (prime minister?) who introduced new laws and made sure the police enforced them. They began executing *suspected criminals*—without a trial. Within a short time, the mafia and drug dealers moved out or were killed along with other criminals. Then the country settled down, but it remained under strict rules. Drug dealing has a penalty of *hanging* and possession of drugs or paraphernalia receives caning (up to 26 strokes with a cane stick) plus a fine and maybe jail time.

We saw no trash anywhere and no stray dogs and cats. There was a $500 fine for littering or not flushing the (Western) toilets; you pay it *right then* or go to jail. There was no smoking in public places. The law is enforced whether you are young or old.

They have fewer accidents because of well-marked crossings you must use; if you get hit by a car while crossing the street, it is *your* fault. Although Singapore was (probably still is) one of the safest places in the world, they have their talented pick-pockets, too.

We hired taxi cabs to go anywhere not in walking distance. Most drivers were honest, but we met one who had no qualms in cheating rich Americans. Christmas Eve day, our call for a cab to take us to Chinatown brought Peter. We should have suspected he was a con artist when he kept saying he wanted to "be our friend" and stay with us all afternoon. He knew where to get everything we were interested in for the best price. We soon decided he took us places that paid *him* a commission but did not have the best price.

When the price on the meter was $13 and we still had not been to Chinatown, Joe said, "Peter, you are no longer our friend. Take us to Chinatown immediately, or I will report you to the police."

He took us to Chinatown. I don't know how much trouble he would be in if we reported him, but evidently he did.

We wasted so much time with Peter that soon it was time to meet our new Australian friends at the Carlton Hotel restaurant. People were getting off work early for Christmas Eve celebrations. Cabs were full or wouldn't stop. Someone suggested we take the bus and pointed out a bus stop. He failed to specify we needed a RED bus. The first bus that stopped was blue. We didn't realize it was going in the wrong direction.

Joe asked the woman driver how far to Carlton hotel. She smiled and pointed ahead. Joe asked several passengers around us; they just smiled and nodded. Each time the bus stopped, Joe asked the driver the same question and got the same response. After 20 minutes the bus was almost empty. She smiled at us and pointed to a building ahead.

It wasn't the hotel. When Joe asked if the hotel was around the corner, she smiled and pointed. Since she had no idea what we asked, she took us to the American Club. Wasn't that where Americans go?

We had found a similar reaction in Thailand. If someone didn't understand us at a restaurant, they smiled and nodded and gave you what they think you wanted. If we tried to correct them, they smiled and said "no have." I never knew if they meant *they* didn't have it or *we* couldn't have it.

At the American Club, we were able to get a cab, but by the time we got to the Carlton, we were a half-hour late. Our friends were patiently waiting. One challenge in traveling without an English-speaking guide is that you never know where you will end up. I began to understand Lee's "Mi pen ri" meaning of it doesn't matter.

On Christmas Eve, all the hotels had parties, but we decided to attend the Orchard Street public festivities. At 7:00 p.m., they closed the street except to pedestrians. The buildings were decorated with light displays reminding me of Times Square on New Year's Eve. People laughed, smiled, and wished each other good luck no matter who you were. We joined the mass of humanity moving in slow waves from one end of the street and back again. At certain points, everyone became quiet while we watched a street performance by a group of children. Between entertainments, someone began singing and everyone joined in. At the far end of Orchard Street, young people were dancing in the street.

When we were back at our hotel, we found Santa had left a stocking on our pillow filled with candies, little cookies, and an orange shipped in from California. What I expected to be a lonely Christmas away from family was an unforgettable one in a place where Christianity is a minority religion.

In May 1994, we flew to London for a bus tour of England, Ireland, Scotland, and Wales. Bus trips allow you to learn more of the history from informed guides. We saw many famous places and took our picture beside them to prove we were there.

We went to the museum where I thought my grandfather's cuckoo clocks had been donated. When I inquired and said how long ago they were donated, I was told they get thousands of donations and cannot use them all. They were probably displayed at some time and replaced with something else. Chances are they were sitting in storage boxes in the basement. It would take lots of research to find out. I was disappointed but understood. I wonder if my grandfather would have.

We went to Bristol where Joe's family came from but didn't have the required time to find the actual land since it was sold and then resold. Like the clocks, it was our family's lost past. It was not as important to us as we had thought.

Our official tour started in Ireland. What impressed me were green rolling hills dotted with sheep and scattered houses. The land was broken into various shaped plots of land with property lines defined by rock walls or Hawthorne hedges. It seems an old rule required a farmer to cut up his land, giving each son an equal share. After a few generations, farms became nothing but a small, useless lot, showing the folly of one generation trying to predict what is best for their descendants.

My memory of Ireland is old castles and destroyed fortifications. The castle at Blarney holds the famous Blarney stone that can only be reached by climbing very steep spiral stairs. Someone holds your legs while you reach down and kiss the stone to have good luck. It requires agile knees and the daredevil attitude of youth or a very strong belief in luck. We had neither.

Scotland Highlands turn from gentle, picturesque hills into majestic mountains called the Moors. Blair Palace is a beautiful white palace set in the green hills. On a sunny day, it shines like a diamond.

In May 1997, we went with a group of Escapees members on an RV caravan through Switzerland, Norway, Holland, Belgium, and Germany. It was the first time we had ever traveled in a caravan.

Germany was my favorite country. I recall rolling hills of grape farms, castles nestled into distant forests, and quaint villages along the Rhine River and valley. One of the distinguishing features of communist East Germany, when the dividing wall was torn down, was the lack of color in their buildings, both in city and countryside. In West Germany, barns and houses are painted in colorful reds and greens.

Side streets in towns are so narrow people park two wheels (or the entire auto) on the sidewalk. However, major highways throughout Germany are well-marked and have services (gas, food, rest areas). You can be fined if someone passes you on your right, so the left lane is used *only* for passing. This makes the traffic flow better.

My memories of Germany include trying to order food in a restaurant where none of the staff spoke any English. It was also a bar and the waiter couldn't figure out our hand signals asking for a menu. A gentleman at another table told him we wanted a menu and when it came, the customer came over and read the menu in English to us. How thoughtful. We have found that people are kind to strangers in all the countries we've visited.

At another restaurant, the waitress spoke a little English and brought us an English menu. They were serving lunch. Most people order beer as a beverage, but I ordered milk. When it came, the milk was hot and in a bowl instead of a glass. It came with a little bowl of sugar and a tiny spoon. Joe said, "I don't think you understand. She wants a glass of *cold* milk to drink."

We found many places around the world don't have iced drinks. Asking for ice brings stares and probably one small cube, which is added to your bill. Ask for water, and they send you to the bathroom. Except at hotels used by Americans, beer or wine is the beverage. (More places now have cola.) The only way to get decaffeinated coffee (which they seldom have) is to ask for "coffee without coffee."

In the Black Forest Hills in Wolfack, Germany, was a recreation of an old-time country house with a circular passageway for protection. The house had a stable, a workshop for repairing harnesses, etc., and a kitchen downstairs. Bedrooms were upstairs and contained chamber pots instead of toilets, plus a large basin and a pitcher of water for bathing. There was no electricity. People went to bed when it got dark and got up at dawn. Black smoke stains and a lingering rank smell surrounded the fireplace.

Next was the world-famous village of a thousand clocks. Many were cuckoo clocks made by the new generation of craftsmen who learned from their fathers. Because they had movable characters and a pop-out cuckoo like my grandfather made, I purchased one.

The last impressive place was Oberammergau, which is a famous village of woodcarvers. (Also famous for production of the Passion Play every 10 years.) We went into many shops and saw the same religious carvings or mountain-climbing men. Someone suggested the older woodcarvers on a side street had better prices. We met a woodcarver who gave Joe some carving tips. His mountain climber was more detailed than any we had seen, so we bought his.

My overall impression of Europe is that too much money, taken from poor working people, was spent on elaborate cathedrals and castles.

Our trip to China during the summer of 2001 was made up entirely of Escapees and their friends. This Orient tour included planes, buses, and a river boat trip on the Great Yangtze River. The Chinese government trains their tour guides; we were not allowed to go anywhere without the assigned guide.

When someone asked an embarrassing question, the guide avoided answering, was vague, or made up an obvious lie. We saw beautiful gardens and parks in several cities, which all the guides called "The People's Park." The explanation never varied. "At one time it belonged to a wealthy family who went away during the night, leaving everything behind. When they didn't come back, we knew they no longer wanted it, so the people took it." How dumb do they think Americans are? Or, do the guides actually believe that?

Shanghai is a modern industrial city where clothing and shoe industries flourish. Silkworms are raised near here providing the silk that is woven into fabrics. Besides being the biggest transportation center and, because of its many (39 in 2001) universities, it is called "the window to other cultures."

Its population of about 13 million permanent residents (not counting tourists) means we saw construction going on everywhere. If we were to go to Shanghai today, we would find a city with model engineering feats that would convince us it was a different place. I'm glad we saw the old Shanghai, even though the modern one is more beautiful.

I don't know if the traffic situation is corrected in Shanghai, but when we were there it was horrendous with more bicycles than cars. We saw people riding bicycles while carrying everything imaginable balanced on their heads or attached to their bodies with ropes: construction supplies, food piled high in baskets, piles of uncovered meat, and one man had *live* chickens and ducks hanging by their feet from parts of his bike. Their heads were all turned in the direction they were going.

Often there was a passenger sitting backwards and carrying part of the load while the driver steered between cars, buses, and pedestrians competing for the right of way. It was worse than driving in New York with Camel.

We visited Shanghai's Children's Palace, where the brightest school children performed for us. Joe got bored and went outside to sit on a bench. A seven-year-old boy sat down saying he wanted to practice his English. The conversation turned quickly to hobbies. The boy said, "I collect coins from other countries. I don't have anything from America."

Joe suspected it was a lie, but reached in his pocket and pulled out some change. The boy carefully picked one of each kind and then said, "I collect paper money from other countries, too."

Children learn early to become beggars, but this boy was already a professional. Adults pursued us relentlessly and were very aggressive. Once I stopped to look at a fan and was swamped by women each trying to sell me their fan. Joe grabbed a fan, handed her some money and pushed the others away fearing I'd be knocked down. Roaming street vendors were demanding and rude. If you bought something from one, others descended on you like vultures. I think they saw rich foreigners as the enemy.

After that, we looked only at souvenirs from vendors who had established stands because they bought a license to sell and were persistent but not aggressive. Sidewalk shops are like little garages, complete with a garage door they close at night. Many sidewalk vendors live behind these doors or in a small apartment above it.

Many deformed cripples sat on the street or curb begging while people stared at them. We learned that, at the end of the day, they were collected and taken home to family or to a shelter until the next day when they were returned to their designated spot to beg from new tourists who felt sorry for them. I felt sorry for them but not because they were deformed. I think they were being used in a humiliating way for other people's gain.

Old buildings are demolished brick by brick from the top down.

When we were there, old buildings were being demolished with sledge hammers, one brick at a time. They started from the top down. Frequently, workmen live in the lower floor apartments (as indicated by laundry hanging from lower windows to dry) until they reach that floor. Workmen balanced on top of the wall as they demolished it. We saw no safety belts for either those demolishing buildings or those standing on flimsy bamboo scaffolding to rebuild from the bottom up. Each brick is cleaned and used multiple times.

The five-day Yangtze River cruise started in Wuhan, so we flew there after two days in Shanghai. Along the often narrow, cliff-bound course of the river, we passed misty mountains, spectacular gorges, bamboo groves, serene lagoons, ancient tombs, temples, and lots of construction at the top of the hills.

Along the riverbanks, farmers were planting crops, and people in the towns continued life as if nothing was changing. Did they realize their farms and towns would be underwater when the project going on around them was completed?

The Three Gorges Dam project began in 1993, and now that it is finished it provides electric power equivalent to 15 nuclear power plants. The dam has ended the past devastating floods. They claim Three Gorges has attracted even more tourists, but the things we saw no longer exist. Whole towns were rebuilt on top of the cliffs to replace the thousand towns and villages now buried deep under water. The dollar cost is believed to be $100 billion, but nobody mentions the cost of human lives.

One day we docked so passengers could take a water tour to see a village. I did not go because it was a small boat, the water looked rough, and I was told there were several hundred steps to be climbed. It turned out to be a wise decision and an opportunity to witness a human conveyor belt. That's the best description I can think of. Joe had the camera so I was unable to take pictures, but because of the distance, I doubt pictures would interpret what I saw.

There must have been 150 coolie laborers, each with the traditional bamboo pole across his shoulders and a basket hanging from each end of the pole. On the side of the hill was a huge pile of coal. The laborers ran in a single line to the top of the pile, where their baskets were quickly filled by men assigned to that job. They then turned in formation and walked down the hill to the edge of the river where, still in line, they dumped their load of coal on the barge. No one broke step in the hour I watched. Like a continuous factory conveyer belt, they went up and down like robots. I wonder how much they were paid and how often they got a break to drink water or eat a bowl of rice. I guess it proves you can keep on going long after you think you can't.

I learned at that time (perhaps it is still that way?), people did not choose their occupation. They were told, "You will be a farmer and sow rice; you will be a fisherman; you will be a teacher, or nurse, or office clerk."

Those in rural areas were not allowed to go into the cities without a special permit. Some of the rural people had never been further than the nearest town. Some didn't realize there was a difference between city life and their own.

In Chairman Mao's (communist) reign, education was belittled and university students sent to remote mountain villages to be "re-educated." Many of the brightest people spent the rest of their lives doing peasant work. Chairman Mao's theory was, "If people do not know a better life exists, they won't try to better themselves." Lack of knowledge has always been a way power-driven leaders keep the population under control.

The Yangtze River was the sewerage disposal of dozens of towns along its banks. Yet I saw women washing clothes on rocks on the river banks. I saw kids swimming in that same water. The sampans and small tour boats that ran up and down the river had a "toilet" but no holding tank, so all human waste went directly into the same water. We had a bathroom in each cabin on our cruise ship with a flush toilet. Where did that end up when we flushed the toilet? I did not have to ask.

We were given one bottle of water, beer, or Coke with our meals, which were included in the tour package. We were warned *never* to drink from the water tap. We could buy extra drinks, including bottled water. When we asked what kind of vegetable or meat we were eating, the answer was either Chinese vegetable or Chinese meat.

Towards the end of our cruise, we had a delicious fish dinner. Our waiter asked how we liked it, and everybody bragged on it. Then he said, "The crew caught them right off the deck this afternoon while you were ashore on a tour."

That stopped us from eating the fish soup served at the *end* of every dinner. It also made me wonder where they got the water they washed our salad vegetables in. For the rest of the cruise, I ate watermelon, which was served at all three meals along with boiled eggs and boiled rice. (Years later, I learned the rice had large quantities of arsenic in it.)

There is so much to see in China, but our exploring was limited to a tour schedule and the polite but strict guides assigned to us. One impression of the cities was the amount of air pollution. Many pedestrians wore surgical masks over their mouth and nose. This wasn't just in Shanghai. It was in Beijing, the capital of China, with so many new, modern buildings. It is an example of how both modern and ancient China existed then; I am told most of the places we visited have been modernized.

Clay soldiers were found in a burial pit in Xi'an.

One place that really impressed me was Xi'an (pronounced CHE-awn). Its history goes back over 3,000 years, but its true fame dates back only to

1974 when local farmers, digging a well, discovered the pit where life-size clay warriors, horses, and chariots were buried. During the following years, archaeologists removed the surrounding dirt and discovered over 6,000 life-size soldiers lined up like an army battalion.

Each face had different characteristics. In the same pit, also made of clay, were 100 chariots and 600 horses. Each figure was painstakingly made of the same clay that gave them the name terra cotta warriors. (Terra cotta means clay from the earth). Some warriors have a broken arm or are missing a head as a result of the excavation. They are more than 2,200 years old. One cannot even guess how many artisans and how many hours were spent creating this army that was buried with an emperor so he would be protected in the next life. I was glad they didn't bury real men and horses.

In one emperor's tomb, they buried 300 beautiful young virgins. These "privileged" girls were real people—not clay figures. They were dressed in the finest silks, had their faces painted and hair fixed to make them as attractive as possible. Then they were given mercury to drink before they were buried in one large tomb with the emperor. Why? To bring him pleasure in his next life. Countless tombs that have not been opened hold other "treasures" buried under the massive mounds of dirt that dot the landscape in the same area where the terra cotta pit was found.

The last place we went was Beijing, where we would see for ourselves the Great Wall of China. It starts at the bank of the Yalujiang River and runs westward through Shanghai Pass and Jiayu Pass and ends at the foot of the Qilian Mountains for a total length of over 3,000 miles. The original wall was built in 290 B.C. but was rebuilt more than 600 years ago during the Ming Dynasty. No picture can truly portray it. The wall follows a narrow ridge of barren hills that stretch further than the eye can see. We climbed steps to one of the lookout towers built on the wall. We could have walked along the wall as far as we wanted, but the steeper grades were not for us.

A street peddler attached herself "to help me." She carried a big bag filled with Great Wall memorabilia and kept offering postcards, books, T-shirts, and did not accept "no." Finally, she disappeared but was immediately replaced with another woman even more persistently trying to convince me she needed my money more than I did. I suppose they needed my money, but to me they degraded an awesome experience.

The immensity of the wall made me wonder how many millions of people spent their lives building this so emperors could feel safe from attack by their northern Mongolia neighbors. The power that a few people through history have had over the people they rule makes me marvel at why there have been so few uprisings. Why do so many people simply go along with the way things are?

The number of vehicles on the roads and the recent construction of bridges makes it seem that many people are doing better than when we were there. Yet history shows that when a nation does well, its people desire a greater piece of the pie. This also can cause internal strife.

One trip I haven't mentioned was to Turkey and Egypt. Since Egypt is in Africa, I will cover it in Chapter 19. We were traveling with Escapees friends Ron and Nancy Hamm, whom we met on the European caravan trip. I looked forward to Turkey because previous visitors told us how friendly they were to Americans. Times had changed, apparently. We did not find it that way. Merchants short-changed us. Nancy understood the money exchange; when one of us received the wrong change, she spoke up. The merchant corrected the mistake without question or apology.

Our tour started in modernized Istanbul. When we got to the rural areas, it was different. In the countryside, huge fields were unbroken by fence lines or farm houses. Then we would come to a village where the famers lived side by side in small adobe homes or apartment buildings.

We saw very few automobiles but many tractors. We watched whole families riding on the tractor or in a cart pulled by a donkey on their way to work the fields. Many men wore western clothes, but women and girls wore long dresses—grey, black, dark blue, or brown—with long sleeves and traditional head covering (not burka—faces were uncovered) even when working in the field.

In Turkey, we saw ruins of ancient historical places, the most famous of which was the stadium in Aphrodisias that dates back to the Roman Empire. This is where entertainments were held. Some were chariot races, but this is also where slaves and captured Christians were forced to fight lions until the lions ate them. Great entertainment? The uncovered ruins showed they had toilets, a kitchen area, and shops where people sold things.

Most remarkable to me was that in Turkey they have uncovered how one civilization was built on top of another that had been destroyed either by a force of nature or by a war. It made me realize if a great catastrophe destroys civilization as we know it, in time a new one will take its place, just as we rebuild a city after an earthquake or tsunami destroys the former one.

Today both Hiroshima and Nagasaki, totally destroyed in World War II by atom bombs, are modern industry cities. There are very few people who still suffer from the radiation effects. It is encouraging to know that even after such total destruction, our planet heals and life goes on.

Troy really was a city, but evidence shows the Trojan horse never happened. It was a story told by a blind storyteller. In Troy, women wearing traditional clothing sat or stood talking to one another, and men did the same. Women walked arm in arm with women; men did the same with other men. I only saw a few teen-age boys and girls walking arm in arm or holding hands.

There were many countries we never got to, but we saw enough to have some understanding of the world we live in and to make me appreciate that I was born an American.

CHAPTER 19

African Adventures (1994; 1996; 2000)

THIS ADVENTURE BEGAN with a letter from strangers in a country on the other side of the world. An Afrikaner couple from Pretoria, South Africa, wrote to the RV industry to inquire how they could go about renting an RV from an individual to travel in the United States for four months. Rental agencies are high and rigs undependable. They didn't know how to answer, so they sent it to Escapees. The letter ended up on Joe's desk.

He saw an opportunity for exchanging vehicles and offered our motor home for four months that summer (their winter). Then, next winter we would fly to Africa and use their travel trailer and tow car.

Pierre and Marjorie arrived in Houston for the start of what became an enduring friendship. After a week of staying on our property to get over jet lag, learn how the motor home operated, and receive some lessons from Joe about driving on the "right side of the road," we showed them Houston. Then they took off on their own.

Eight months later, they met us at the Johannesburg airport in the Republic of South Africa, and we began two months of an adventure that was a highlight of my life. We originally planned to stay for three months, but our timing was bad. Pierre worried about our safety and believed we should leave Africa before the upcoming election ended apartheid.

One has to know a little of South Africa's history to understand the turmoil that started in the 1800s and continued until after World War II. The two white races, the English and Afrikaners (also known as Boers), had fought each other in two bloody wars to gain land and political power. Both groups thought the white man was superior to the black.

The two largest African races in South Africa—Zulu and Xhosa—had also been enemies forever. In 1923, the African National Congress (ANC) was officially created to unite black Africans. At the conclusion of World War II, Nelson Mandela took the ANC in what appeared to white people to be the collapse of their culture. Since the National party (all white) was in power, Mandela, along with many other blacks, was thrown in prison as a political enemy. Apartheid became the white man's weapon for keeping the black races from obtaining power.

When we were there in early 1994, Mandela had already been released after 27 years of cruel imprisonment that made him the hero of black Africans from all the tribes, as well as the colored people.

The coloreds, consisting of Indians, Asians, and anyone who had intermarried within any of the races, also wanted the power to be taken from both the English and Afrikaners. Mandela hoped Africans, who were finally allowed to vote, could take over the government in this upcoming election.

Our friends, like most Afrikaners, were of Dutch, German, and French descent. Their ancestors had settled in southern Africa many generations ago and made up their own language. They were fundamental (Old Testament) Christians whose religion believed that God intended black people to be slaves. They were afraid of what would happen to their country if the Africans came to power.

They had reason for their fear. When the Africans took over the Rhodesia government, they immediately changed the name of the country to Zimbabwe. Many Boer farmers were killed by their former laborers and by new government officials who wanted to own their farms. The white farmers who were not killed were forced to flee with the clothes on their backs. They took with them the knowledge the blacks needed for running farms and businesses.

When Robert Mugabe became president in 1987, the true downward spiral began through corruption and mismanagement. Every time people thought things couldn't get worse, they did. First, the farms stopped producing, and then the businesses had only foreign-made products to sell. This turning of the tide took time. When we were there in 1994, Zimbabwe was already depending primarily on tourist trade. By 2000, very few crops were being grown; today the old farms are mostly slum areas and poverty has become even more widespread.

Our friends saw a disturbing trend in their own country. Even before the election that would forever end apartheid, many blacks were taking advantage of their uneducated black brothers who would believe anything if it was a black person who told them.

We talked to people who thought that when Mandela got in power, all maids and laborers would become the home owners, and the white people they had worked for would now work for them. Some had paid money every month for as much as three years to these black "friends" so they could have their pick of a house.

One black lady said, "After the revolution, we will get one of those plastic cards that white people take to the bank, and we'll come home with as much money as we want." When we asked where the money came from, she said, "The government."

It seemed many believed white people had some kind of magic and would be forced to give the magic to the black people. They had not come much further in understanding than the time when a young girl in one black tribe had a vision and convinced everyone to destroy all their food and kill all their cattle.

I thought about the black people in America. Some are dissatisfied and feel they don't have equal rights, but when you see them sitting around in groups, they are laughing and talking with each other. They have a good time. The blacks we saw in South Africa wore white shirts and nice pants or clean dresses but looked solemn, sullen, and were silent. They seldom laughed even when they were eating together or sitting on a curbstone watching life pass them by.

After Pierre and Marjorie picked us up at the airport, we went to their house, where we quickly discovered just how seriously they took the anticipated change. They had a 10-foot block wall surrounding their property, with an iron gate that could only be opened with a special key. Two vicious looking Rhodesian reds (gentle pets but bred to attack blacks) roamed the premises. When we got to the front door, I saw attractive iron grill-work that was meant to keep intruders out, with another solid door behind it. Both doors were kept locked at all times.

The bedroom they had set aside for us had iron grillwork, stronger than prison bars on the outside of all the windows. At night, action lights in our bedroom came on if we went to the bathroom. The entire house was under similar protection. We learned Pierre never left home, even to go to church or walk the dogs, without a pistol concealed under his sweater or vest. His vehicles had the most sophisticated security systems Joe had ever heard of. To me, this was living captive to fear.

Pierre warned us that when we passed any blacks on the street to stay alert, but don't speak to them or look them in the face. We followed his instructions when we were with him, but both Joe and I smiled at them or said hello when we were alone. The reaction we got was different from what Pierre feared. Most smiled back and said hello, but some looked shocked as they mumbled hello. Only a few looked down at their own feet.

We stayed two weeks in our friends' home, visiting in their relatives' homes. We enjoyed many "braas" (back yard cookouts) and saw the local sights, including the gold mine in Johannesburg, a pioneer museum, and a unique restaurant made of former railroad trains.

The entrance to the restaurant was a replica of an old railway station with platform signs. A lounge car served as a bar for ordering pre-dinner cocktails. There were four separate dining cars decorated in Edwardian style that seated up to 176 people. The four railroad carriages circled a huge buffet area that had every kind of meat, fish, and salads imaginable. In addition to the traditional foods, there was every kind of edible game found in the area, plus a huge dessert table. We never made it that far.

I tried ostrich, crocodile, impala, dik-dik (a small antelope), and warthog (like pork). All were delicious. Feeling brave now, I tried elephant, zebra, and wildebeest, which were okay, but I could never want a full meal of them. Then I tried giraffe, which was too tough to eat. Joe said it tasted like shoe leather. Personally, I've never eaten shoe leather, so I took his word.

Pierre and Marjorie borrowed their son's motor van and caravanned with us the first couple of weeks. We stayed in both national and private campgrounds. South African campgrounds were clean and had great facilities, except there are no sewer hookups because their trailers do not have holding tanks. Like New Zealand, they use porta-potties at night and the campground rest rooms for toilets and showers. You carry your own toilet paper.

Kruger National Park is one of the best-known wildlife reserves and the most visited. You can take a safari closed-bus tour with up to 45 passengers or drive your own car. Hikers can trek with a ranger to certain areas. We went in Pierre's car.

This area was home to herds of all kinds of animals until at the end of the 19th century, an epidemic of rinderpest, carried by the tsetse fly, caused thousands of animals to die. Paul Kruger, president of Transvaal Province, fearing Africa's animals would become extinct, convinced everyone to set aside a parcel of land the size of Massachusetts as a wildlife preserve. We saw a lot of animals, but not up close. Regular cars stay on the roads.

When we returned to the house, we left the trailer and most of our clothes with Pierre. His house remained our home base and mail collection point. At everyone's insistence, we took the famous Blue Train from Johannesburg to Cape Town. From there we would take a bus tour.

The train tickets had to be purchased ahead and were printed in both Afrikaans and English. We read the departure and arrival information and skimmed the rest, intending to read it on the train. That was a mistake. We missed the instructions about dress codes.

Ignorant of what lay ahead, we left Joe's only suit and my dress hanging in our friends' closet and took only a change of casual clothes in one small suitcase. We had learned to travel light and didn't expect to go anywhere that required dress clothes.

Every couple had their own private stateroom with picture windows to watch the scenery. Waiters in white jackets were at our beck and call. There was an elaborate lounge with free drinks next to the dining car, where they served free lunch, dinner, and then breakfast the following morning. The train made one stop at the famous Kimberly Diamond mine for us to admire it and buy souvenirs.

Joe, with his joke-telling, had made friends with the captain. After we were all back onboard, Joe went to the lounge to have a drink and visit with anyone there. The bartender was alone. The captain came by wearing a white tuxedo.

"Why the tux?" Joe asked.

The captain was not sure if Joe was leading up to one of his jokes. Guardedly, he replied, "You do know there is a formal dinner tonight, don't you?" He looked at Joe's short-sleeved red-plaid shirt and said, "You do have a suit and tie, don't you?"

When Joe shook his head, the captain rolled his eyes. "I can lend you a tie." Then, as if in afterthought, he said, "You do have another shirt, don't you?"

"Yes. It's just like this one except it is *blue* plaid."

The captain looked as if he would faint. "Did you read the information on the back of your tickets where it said 'elegant dress is required for dinner'?" Again, Joe shook his head. "What about your wife? Does she have an evening gown?"

"No," Joe said. "She has long pants and a couple of blouses. Does that mean we won't get any dinner?"

The captain hesitated, obviously tempted. We had paid for this special seven-course dinner. He had an idea. "Can you come half an hour before the dining car opens?"

Joe agreed. I put on my best slacks and blouse and he changed into his blue-plaid shirt and put on the borrowed captain's tie. The only pair of long pants he had with him were brown. It was that or the grey shorts he had been wearing, but they were even less appropriate.

They had someone at the door watching for us, and the moment we arrived, they whisked us down to the end of the dining car and put us in a little corner that was curtained off from the rest of the seats. We heard people coming in and people chatting as they found their seats. Joe was nursing a drink and I had a Coke which seemed to be in low supply. I think I was the only one onboard not drinking wine or fancy drinks.

It was an elegant dinner. A number of knives, forks, and spoons were set in a particular order on the sides of the plate and above it. We had no idea what each one was for. We seldom do formal dinners, and when we do, we wait to see what other people pick up before we select ours. This time we could not see what anyone else was doing.

To amuse himself, Joe rearranged his silverware after every course. Each time the waiter came to bring us something or pick up an empty dish, he patiently put the remaining silverware where it belonged but never said a word.

Finally, that long, long dinner was over and we wanted to go to bed. We were supposed to wait until all the other guests had left. We heard them ordering more drinks, and the smoke from cigarettes and cigars began penetrating even our secluded corner. It looked like a very long evening, and Joe no longer had any silverware left to amuse himself with. We toughed it out for what seemed like an eternity until we heard people ordering even more drinks. Joe said, "This is ridiculous (or words to that effect). Let's go to our stateroom."

An amazing thing happened when we stepped out from behind the curtains and into the aisle. All chatter slowed, and in seconds a deadly silence took over. Ladies in formal evening gowns offset with elaborate jewelry and men in tuxedoes stared as we paraded up the aisle past their tables. When we reached the end, Joe turned back and said in a loud voice, "Sure was a great dinner, wasn't it? We'll never forget this evening."

As soon as we were out of sight, it was as if someone suddenly turned a water faucet on, and a roar of chatter erupted behind us. We broke down laughing. No sense being embarrassed now. When we reached our stateroom, we found the seats were already made into beds. There would be no interruptions, but we still had breakfast to do.

Although we'd never see these people again, we felt we had disgraced our country because it was obvious we were the only Americans. Since there was no way to undo that, we decided to wear the same clothes as the night before, except Joe didn't wear the tie.

We greeted everyone as we passed their tables. Most were too busy looking at their menus to respond to our greeting. I told myself that people who live in a black-and-white world, ignoring all the grey areas, miss the richness of the experiences we have had. Our friends were right, the Blue Train ride was special, but not for the reasons they thought.

Our time at the Cape was divided between walking around Table Bay and taking a Cape Peninsula tour with one other couple, who spoke only a little broken English. Our guide was an interesting colored lady, a mixture of Indian and Dutch and perhaps something else.

She was definitely looking forward to ANC taking over the country and "making things more better for blacks and coloreds." Both she and her husband were teachers. Her university training was not in subjects she wanted, but it was her only way to get into a *white* university at that time. I'm sure after the blacks took control that both she and her husband had much better teaching jobs and bigger paychecks.

Our first guide on the garden tour was part American Indian and a citizen of both the U.S. and South Africa. He took us on side trips to scenes of particular beauty that were not on his agenda. It is this sort of thing, plus being able to ask questions, that make a tour worth the investment, but it all depends on the tour guide. Some have a lot of knowledge and are happy to share it; others give you only the canned version they were taught.

On the second day, we had a couple from Austria as traveling companions. This tour included an ostrich farm, where we learned that ostriches don't really bury their heads in the sand. They stretch their long necks out and press their heads flat against the ground, so that from a distance it looks as if their heads are buried. They can give vicious bites with their strong beaks and attack you with their claw feet, so if an ostrich starts running toward you, your best defense is to drop to the ground and lie flat and still.

At the end of the ostrich farm tour, we were given a display of their racing ability. After the race, everybody was invited to ride one. Out of our group of 24 travelers, Joe and I were the only ones who accepted the challenge. (We were also the only Americans.)

We were instructed to hold onto the ostrich wings, but as soon as my ostrich moved forward, I slid off his back. He kept running to the finish line without me. Because Joe went all the way, he was declared the winner. I didn't get a picture because I forgot to turn on the camera. I guess that's why Joe usually takes the pictures.

What I didn't understand was why no one else was willing to get on the back of an ostrich—even if, like me, they fell off. How many chances in their lives would our tour buddies have that opportunity again? How many, like Joe, could brag to friends that they rode all the way to the finish line? You have to be willing to risk looking foolish or being unsuccessful to enjoy all the opportunities travel offers.

I stood on an ostrich egg and was amazed at how tough they are. They say, one ostrich egg is equivalent to 25 hen eggs. I purchased an ostrich egg shell that had a hand-painted leopard sketched on its glossy, pitted surface.

Kay inside cage petting three-year-old cheetah.

We stopped at Cango (limestone) Caves (a small version of Carlsbad Caverns) and then went to a crocodile farm where we learned more about them. They also had a series of cheetah cages with injured and orphaned cheetahs. Care of the animals was paid for by tourists willing to pay $50 to go into the cheetah cage (with an employee) and pet a three-year-old cheetah. The money went toward saving more injured or orphaned cheetahs and crocodiles.

That was an opportunity I couldn't miss. I went with two other women tourists.

Before we made our way through a couple of cages to the one where our cheetah lived, we were told not to step on any tails. It can make a cheetah mad. When we got to our cheetah, we were allowed to rub the fur on her chest. While I did that, as a sign of mutual affection, she licked my arm with her sandpaper tongue. Then I stuck my thumb in her mouth and she sucked on it. What a memorable experience! To me, life is about weighing the risk against your desire. Wild animals are undependable, but the law of averages is on your side if you do what the guide tells you.

(I recently heard wild animal back yard "zoos" have arisen in parts of America so people can pay to have their picture taken with a tiger or chimpanzee. If you decide to do this, find out if the guides are trained and how much experience they have. Remember, there is always a risk when wild animals are involved.)

There were many interesting stops between Port Elizabeth and Durban, where we had a hotel room on the 25th floor overlooking the ocean, beaches, and city. It was beautiful both by day and night. From there we drove through many sugar cane fields and past many Zulu village settlements on our way to Shakaland, a reconstructed village built as a movie set and then turned into a tourist attraction. Africans, dressed in native costumes, demonstrated their food making and craft skills. Joe used a lot of our film taking pictures. When they were developed later, I was not surprised to find most were the beautiful young topless girls.

The next day we returned to Pretoria. We had one more 10-day trip ahead of us to Zimbabwe. This country is north of the Republic of South Africa and south of Zambia. It is best known for being the location of the world-famous Victoria Falls.

The animals were why we came to Africa, so again we left most of our clothes and the souvenirs with our friends and took off with two small suitcases. We flew to Harare airport in Zimbabwe, and then took a smaller aircraft to Hwange National Park and Lake Kariba, where our first safari started. Lake Kariba is a water wilderness lake, so huge that it took an hour to get across it in a *speedboat at full throttle*. It was sunset and the lake was calm; it was a fantastic view.

When we arrived at a small, isolated dock, daylight had faded, and the guides met us with kerosene lamps to light our way up the hill's many steps to Sanyati Lodge, where we would stay for the next three nights. The lodge was actually a log hut with a thatched roof. Inside, the twin beds were enshrouded in mosquito netting, and we had a private toilet and shower.

"When you are ready, continue up the trail steps to the big lodge at the top of the hill. That's where meals take place," the guide said.

We found these meals, served on a big circular table, were a source of great enjoyment because we talked with visitors from England, Holland, Germany, and Australia. We discussed many things and compared our lifestyles during three-hour meals where liquor flowed freely and wine glasses were quickly refilled. Getting a plain Coke was more difficult. Pre-dinner drinks started at eight, and dinner was finally over at eleven p.m.

At Sanyati, animal viewing was done on small boats and consisted mostly of looking for animals coming out of the Bushveld to drink. We saw mostly hippopotamuses, who spend their waking hours in the lake with only their eyes and noses above water. It was a great place to observe birds, too. This was not as enchanting as the brochure indicated, and after one morning of riding in a small sightseeing boat on choppy waves, I stayed at the lodge whenever the wind whipped the waves, which was most of the time. Joe went on every trip but said he didn't see much except hippos.

It was a restful, tranquil place, and I made friends with some monkeys who came to my hut for a handout. After the first day, I carried something back from breakfast and lunch to share with them. They ate anything, grabbing the food from my hand. This was so different from the gentle fingers of kangaroos.

On the fourth morning, we again took the speedboat across the lake and then flew to the other side of Zimbabwe, where we were met by a guide and taken to a temporary camp for lunch and given an empty bungalow where we could take a nap. Our guide to Camp Makalolo would not be able to pick us up till afternoon.

When our guide arrived, we drove for three hours through Hwange National Park to an isolated tent camp. It was dark when we arrived. As if sent to be our welcoming committee, a small pride of five lions sat right beside the road about a mile from our camp. When the guide turned the headlights on them, they just stared back but never moved. From that introduction till the ride back to civilization, Makalolo was the most fascinating adventure we've had. I wished we hadn't wasted three days on the water safari, but brochures can be deceiving.

Makalolo is a permanent, but primitive, camp. All our meals, including very elaborate dinners, were prepared for us over open fires and on a primitive wood stove. Bread was baked in a clay oven, as were some desserts.

We had our own sleeping tent with two narrow cots, two small windows with screens, and a screen door for better ventilation, plus a small adjoining tent bathroom with a shower, so we didn't have to go outside at night. Lions frequently cut through the camp at night, but they ignore the tents, which are inside a rock wall about three feet high. It was here that our real animal watching took place.

Spiders and mosquitoes invade the tents at night, so we slept under mosquito netting hanging from the tent top and tucked tightly under our mattress. There was one pull-chain light hanging in the middle of the sleeping area and another in the shower/toilet section. We had left our flashlight behind, so we tried not to get up once the light was out. Getting in and out from under mosquito netting was a challenge.

The African Bushveld is one of those magical places you can never understand unless you experience it for yourself. It is a place where countless animals roam freely. We were isolated in a world where animals make the rules, but we were also insulated from the troubles of the world. For a week, we had no news for there were no telephones, no televisions, no radios, and no newspapers to interrupt the tranquility.

There were no alarm clocks either. Time was measured by where the sun was. We were awakened sometime around five o'clock each morning by a native boy standing outside our tent window saying, "knock-knock." That was our signal to go to the big hut.

When Joe and I arrived the first morning, most of the other guests were already gathered around the fire. We were handed a cup of hot coffee and a rusk. Rusk is a small hard-as-a-rock piece of bread that is the African donut. The only way you can eat it is to soften it in hot tea or coffee. We would have a real breakfast when we returned from the dawn "pan ride."

We were divided into groups of six and assigned to a guide and a bright yellow open-topped Land Rover. Our group included a couple from Germany on their honeymoon and a couple from Australia. The wild animals knew yellow vehicles were harmless, so they paid no attention to us. This allowed us to get as close as 15 feet from them.

We went to the watering hole where Henry the hippopotamus lived. Today, there were no other animals there. Our guide told us Henry had arrived at the pond many years ago and stayed even though droughts lowered the water. Now even the deepest part no longer covered his back. In summer, a hippo needs to cool its thin, smooth skin by submerging itself in water.

An adult hippo can be up to 15 feet in length, stands five feet tall, and can open his mouth up to four-feet wide. Its ears are directly above its eyes, and it mostly keeps them above water. It can totally submerge its body, walking on the bottom of a lake or river, for as long as five minutes without lifting its head to breathe. This pond did not allow that.

Henry spent his adult life in a pond near Camp Makalolo.

Hippos are vegetarians, so they go onto the land at night, sometimes walking miles to find a preferred-type grass. Their short, stumpy legs make them appear clumsy, but they can run 18 mph and turn on a dime. They can climb but avoid steep banks because they cannot go back down steep inclines.

If challenged in any way, a hippo will attack and kill people either in the water or on land. They are affectionate with each other and like resting their heads on others in their family group of 10 to 30 hippos. Each family is territorial and has a male leader who fought for and won that position. He is the only male in his group that can breed females.

Then why was Henry the only hippo in the pond? Males are welcome to remain in the group until they reach adulthood and challenge the ruling male for the right to breed. That is probably what Henry did, but he challenged a patriarch that was bigger and stronger than he was. He should have backed off when the patriarch started yawning. That is hippo language for "Don't mess with me."

When they fight, each hippo swings its massive head from side to side trying to hammer the opponent. If they use their teeth, it frequently causes permanent injuries that end up with one or both dying. To stop the attack, the losing hippo submits by turning around and "tail-paddling" to spread his feces. Then he slowly walks away, hoping his wounds don't get infected.

No one witnessed Henry's battle, but those who understand hippo behavior say that when a hippo is defeated, he is usually banished from the family.

He then wanders by himself until he finds a group that will accept him, usually a herd of other defeated bachelors. In Henry's case, he decided to stay by himself in this shallow pond close to Camp Makalola.

We visited Henry every morning and evening on our way to various animal viewings. Often there were other animals drinking there, but Henry stayed away from them. A guide who had been there for years told us that one day a female wandered by and decided to stay. Henry seemed happy to have someone to hang out with. The guides named her Henrietta. The couple managed to mate underwater, as is the custom, and eight months later she had her baby, Henry's only son. I hope his son grew up to be more courageous than his father.

In most cases, the mother hippo isolates herself when she is about to give birth, but this pond was so shallow that giving birth underwater must have been difficult. As soon as it is born, the baby floats up to the top. Normally, the mother and baby will stay isolated for 44 days before rejoining the herd. Henrietta and Henry stayed together while the baby grew strong. Most baby hippos are prey for crocodiles, leopards, and lions until they are about a year old. The predators would stay away from two adult hippo guards.

For a reason only hippos know, one morning at sunrise the guides saw the forms of Henry, Henrietta, and her son walking away. Henrietta may have instinctively realized this shallow pond was not a good place to raise a family. Or she may have wanted more social life than was here.

Several days later Henry returned alone. No one ever knew what happened. Some guessed another male had wandered by and caught Henrietta's eye. Hippos are very fickle. Henry had no further scars, which indicated he had not fought to keep her. Perhaps he was afraid of getting hurt. I could understand that. Whatever the reason, Henry returned alone to the only security he knew.

He taught me a valuable lesson: *Fear keeps you from enjoying all the adventures life offers.*

That was proven the second night. Something woke me. I bolted upright, and in the inky blackness, something grabbed at my hair and covered my face. I heard a lion roar, and, in panic, I struggled to free myself until I realized it was just the mosquito netting. Another roar came and this one sounded very close.

From his cot on the other side of the tent, Joe said quietly, "There's a lion prowling around outside our tent. Go back to sleep." I listened to the lions calling each other and baboons chattering their warning before I fell asleep. I told myself we were safe inside our tent.

The next morning on our way to the big hut, we paused to look at the ground outside our tent. Sure enough! There was the spoor, the footprints, of a lion as close as 15 feet from our tent. Everyone was talking about the lion and the different guest reactions. One older lady who arrived yesterday got up, dressed, packed her suitcase and sat on the bed until her knock-knock came. She demanded to be taken back to civilization immediately.

This disrupted the routine. One couple who were due to leave after the dawn drive had their final drive shortened while the hysterical older woman waited impatiently. Then the guide took them and their picnic breakfast on the three-hour drive. I'll bet it was a long three hours. I wonder if that woman ever cared that she cheated two people out of the last morning they had paid for.

The German bride told us that when the lion started roaring it sounded as if it was right outside their tent. Her husband, Frank, jumped out of bed, ran across the room, and insisted on sleeping with her "to protect her." The cots were never intended for two people.

Later that same morning, our guide spotted an elephant too deep in the bushes to really see. He stopped and gave a lecture on elephants. The African elephant is different from the Asian elephant, which can be trained to entertain and to work. No one has ever tamed an African elephant and probably never will. African elephants are living a twilight existence, trapped in a narrow strip of grasslands and brush that is becoming a little smaller every year.

One reason is their voracious appetite. Elephants eat enormous amounts of food and drink up to 20 gallons of water a day. Worse still, they often uproot an entire tree to get the tender leaves at the top. You can follow an elephant's tracks from the destruction of the brush as well as their spoor and droppings on the ground.

Another reason elephants are in danger of extinction is that poachers have slaughtered them by the hundreds for their ivory tusks and the feet, leaving the remainder to rot or be eaten by vultures. The U.S. no longer allows imported ivory except under specific conditions and tariffs, but other countries buy the tusks for carvings and the feet for a decorative table for homes in their part of the world.

There have been stories of deaths to tourists. (Marjorie was scared of elephants at Kruger Preserve.) What should you do if an elephant comes walking toward you? If he is holding his ears out from his head and swinging his trunk from side to side as he rocks back and forth, it is a sign of anger. Stand absolutely still. You must fight your natural instinct to run. He may keep you standing there for an hour or more, but probably his anger will eventually subside, and he will turn his back and walk away. Probably?

When an elephant kills a tourist, it is usually caused by the tourist's ignorance or foolishness. Some tourists are so determined to get a photo to take home that they ignore the animal's display of anger, and then he gets *really* mad. Jumping in your automobile provides no protection. He can stomp you and your car into the ground, along with that so-important photograph.

Our guide asked if we wanted to stalk the elephant in the bush. Joe and I grinned. Joe said, "We wouldn't miss it for anything!" The Australian couple also wanted to go. Frank, the German, said they would wait in the Land Rover. Then his wife said, "Not me! I'm going with them."

Frank was distressed. He didn't want to go, but neither did he want to wait in the open-top vehicle by himself. His decision was made when he saw the guide pull out his rifle and check it for bullets. He insisted he and his wife would bring up the rear. Maybe he thought being at the back gave them a better chance to escape.

The guide said, "No talk. Not even whisper. Use hand signal. No make walking noise. Stay close together."

We were lined up ready to go on when the guide lifted a handful of dirt and watched the way the wind blew it. After doing this a couple of times, he seemed satisfied and told us to follow him. I suspected this entire event was play-acting, but it was fun to pretend we were great elephant hunters as we moved silently from tree to tree, slowly getting close to this mighty animal eating breakfast.

The elephant, alerted to something not being right, stopped eating. The guide motioned us to head back. Frank panicked and was about to run when his wife grabbed his shirt, forcing him to stop. Then she led him back over the trail, with the rest of us following. I managed to suppress laughter. I think he picked the right wife.

Our schedule was a long predawn pan drive to various watering holes, where we saw many different animals, return for a leisurely breakfast, a short rest where you could take a nap, sit on the lookout deck and watch the grasslands, or have a drink at the bar. The next pan drive would start around eleven—before lunch. After lunch we'd have another two-hour rest period and then a Bushveld drive before dinner. It was after 10 o'clock before we finished dinner. Then there was an optional late-night drive where we observed different, smaller animals.

One morning Joe decided he wanted a beer, so we went to the big hut where meals were served and the bar was open all day and evening. There was no one there except the bartender, who was doing some paperwork. But at the wide opening at the other side of the big hut, an elephant had his head and one foot inside the hut. "Hey! That's an elephant," Joe said.

The bartender looked up at the elephant and said, "Sure is. Did you want something to drink?" Joe got his beer, and about that time the elephant backed out and wandered away.

Our days at Makalolo passed swiftly. We had seen so many animals—warthogs, giraffe, herds of elephants, zebra, several kinds of antelope, cheetah, leopards, cape buffalo, baboons, jackal, and even a porcupine. The great thing was that we could go to within 400 yards of them and often even closer.

We had one more adventure the morning we left. No one else was leaving that day, so while they were on the regular predawn pan drive, we were by ourselves on a shorter drive with the guide who would take us to the airport. We had gone less than a half mile from camp when the guide suddenly switched off the headlights and eased up on the motor. As he inched his way forward, I stared into the semidarkness that comes before the first light of dawn. Joe punched me in the ribs and pointed ahead. Blocking the road was a pride of lions. As my eyes became accustomed to the semidarkness, I began to count them. "It looks like seven," I whispered to Joe.

"SHHH!" the guide warned.

Joe poked me again and pointed to two strolling in from his left. Then I looked over my side of the Land Rover and stared into the eyes of an approaching lion. She paused and then ignored me. She looked so tame, so innocent, that I had to fight the urge to scratch her back. Her tail brushed the metal side of our vehicle.

Now all the lions were lying down in what seemed an aimless pattern in front of us. I counted again. There were 11. One was a male, at least two were grown females, and the rest were probably teen-agers.

The first glimmer of light was awakening the Bushveld around us. In the distance, I saw the bodies of a huge herd of cape buffalo grazing in the grassland. Near the buffalo was a solitary wildebeest that either strayed or was an outcast from his herd. Unless you are a leopard or a cheetah, the Bushveld is not a place to be all on your own. Large herds of zebra, buffalo, wildebeest, and antelope often adopt an orphan of a different species.

The lions were watching them as intently as we were. Then two lions from one side of the pride got up and started creeping toward the left. Within seconds, two more got up and began sneaking through the grass toward the right. It was hard to keep track of them in the dim light, but each pair stayed fairly close together. Two more got up from each side and two from the front.

In a few moments they began fanning out through the grass. Finally, the male sauntered after them. We could barely spot the various lions as they crept through the tall grass. They seemed to have an established plan, with each knowing his/her part.

Suddenly, one of the buffalo lifted his head, and a loud, ominous bellow came from his throat. Instantly, faster than I can write the words, the buffalo bunched up. Within seconds they were in a ring, shoulder to shoulder with their massive heads displaying deadly horns.

Young and pregnant cows were inside the protective circle. The wildebeest joined the outer circle. We could talk in low voices now, and the guide explained that the lions fanned out in such a way that if the buffalo panicked and broke ranks, running in all directions, the lions would be able to pick off one or two of the calves or pregnant females. The first glow of dawn was lighting the scene when the guide said, "Look!" and pointed to two impatient young lions, who with the daring of youth, began running toward the herd.

Again came that ominous bellow and, as if it was a known command, the buffalo as one mass turned and charged the two young lions. As the herd rushed towards them with massive horns aimed for the kill, the two lions turned tail and raced away so fast that it seemed their feet were not even touching the ground. The threat was over. The lions would not feed this time.

We sat in awed silence for a couple of moments before the guide started the motor. We headed toward the familiar watering hole where we would say good-bye to Henry and see what other animals came to drink before we headed to the airport. All I could think of was the sight I had witnessed—not in a National Geographic film—but here, right in front of me. In an African dawn, I had seen animals working together. Was it instinct, learned behavior, or some mystical signal?

If the buffalo scattered in panic, some of the young and old would have little chance to survive. Together, protected by the strongest, they were safe. People can and often do act alone, depending on their own capabilities. Yet throughout history, our greatest achievements have come when we work together for a common cause, unconcerned with who does the most or who gets the credit. It was a lesson now engraved in my memory.

Five lions greeted us at Makalolo and a pride of 11 bade us farewell. When we arrived at Harare airport, a guide took us to the hotel we had booked at Victoria Falls. The walk around Victoria Falls is impressive, especially one view that showed all three falls at the same time. On the way back from the falls, we passed a native with some crude carvings of animals. He was just finishing a hippo, and I bought it in memory of Henry.

This was an overnight stop only. My impression of the expensive hotel where we stayed was its ongoing war with a tribe of monkeys who long ago realized tourists had food. If tourists ignored the rule to keep their window shut, whether they were home or gone, they could be invaded by monkeys that looked cute but destroyed a lot of things through curiosity or in search of hidden food. Sitting in the garden, watching them climb the walls and try to open windows, was my last memory of Zimbabwe.

When we got back to Johannesburg airport, Pierre and Marjorie were waiting for us. That night at supper, we talked politics with them and their children who came to say good-bye. The political and human problems are so intertwined that there is no happy solution for all concerned. The ANC had caused some eruptions while we were away, and our friends worried about our safety. They were relieved we were leaving the next day.

After we got back to Texas, we learned Mandela was swept into office, and the blacks immediately began taking over government jobs that only whites held before. The changeover was more peaceful than our friends anticipated. How often we worry ourselves sick about things that don't happen.

Eight days after we arrived back in the States, I came down with what we thought was flu. It was not a good time to get sick as we were leaving for the Spring Escapade in Chico, California, a very long drive from Texas. I expected to be over the flu before we got there. We were traveling by RV, so Joe fixed the bed to stay out permanently. I could lie down whenever I wanted, which soon became all the time.

I don't remember how many days it took to get there, but I remember being very sick with chills and fever. I went into a restaurant once, but the smell of food made me too nauseated to eat. Sometimes Joe fixed himself something in the RV; if he craved a real meal, he went in by himself and brought me something I tried unsuccessfully to eat.

When we pulled into the Escapade, everyone became concerned. I stayed in bed that day (Saturday) but got up on Sunday long enough to give my (shortened) welcoming speech. Then someone whisked me away from the crowd that tried to surround me and back to our RV. Bud was trying to find a doctor. By now, we all realized it wasn't flu. There were two possibilities—hepatitis, which is contagious, or malaria that is not.

I don't remember details, but I ended up in the emergency room where they did tests that determined it was malaria. I was taken to a private room where I spent 10 days, missing everything except my welcoming speech. Joe,

Bud, and especially Cathie visited, but no one else was allowed. Yet it seemed every time I opened my eyes, a strange face was staring at me.

I later learned I was known as "the old lady with malaria in room 214." I was the talk of the hospital since we no longer have malaria in this country. The hospital staff was curious to see what someone with malaria looked like. Actually, I just looked old and sick. How disappointing.

Pierre and Marjorie came over in1995 and used our RV one more time before we sold it and bought a smaller unit. At one point, Pierre and Marjorie were going to come to America to live, but after realizing their families could not follow, they decided against it.

It was not until 2000 that we returned to Africa. This time we traveled with our Escapees friends, Nancy and Ron Hamm. The trip included a visit to Egypt, which did not meet my expectations in sightseeing or attitude of people who took every opportunity to cheat us.

First we flew to Abu Simbel, a famous temple in Egypt. Our "English-speaking" guide spoke rapidly and with a heavy accent none of us understood. We were the only English-speaking tourists. Finally, a bus took us to the shuttle plane we had come in on. Security was much tighter than when we came in, and they found Joe had the same pocketknife in his pocket that he carried throughout Turkey.

Joe got into a heated argument with security. Finally, security agreed, if he gave it to them during the flight, he would be able to get it back when we landed in Aswan. Nancy, Ron, and I worried that he was going to be put in jail or heavily fined over a $25 pocketknife. It was his favorite knife that he carried for years. (He put the knife in his checked luggage on future plane trips, but it reminded me of Pat losing his prize horse over a stolen $15 rifle.)

In Aswan, we boarded the small cruise ship that took us down the Nile River to Cairo, stopping to see famous temples along the way. We passed Edfu during that four-day Nile cruise. As we approached the town we heard the echoing wails of various mosques calling Muslims to evening prayer. (They pray six times a day.) The sound of wailing is impossible to describe, but it made me want to put my hands over my ears to shut it out. I resisted the temptation because it would be disrespectful and might land me in jail. We watched the town shut down as people closed their shops and headed for the nearest mosque. One guide told us the architect of a mosque was killed as soon as it was finished so he could not build a replica elsewhere.

I did not believe him at the time, but since then I think it may have been true. Muslims believe dying for a "cause" gets them to heaven with something like 70 young virgins to keep them company. Better than money or fame. The same guide told us one leader invited his political enemies, their armies, and even the common people who believed in them to a free banquet. Once everybody was inside, the doors were locked and everyone was murdered. The only way out was to jump over the sea wall to certain death. Was this a party to end all parties or a story to entertain naïve tourists? Often it was difficult to distinguish truth from fiction.

Our tours were mostly to famous tombs from which all the treasures had been stolen long ago. Of course, each temple and tomb had something unique about it, but old temples with old gods and century-old hieroglyphics on walls did not interest me.

In Cairo, we were taken to the huge bazaar. The alleyways between shops are narrow, smelly, and dirty in places. Most vendors stay inside their shops unless you stop to look at something on display. Then they come outside and start announcing bargains. I knew the price quoted is never the price a bargainer pays, but I find bargaining demeaning. Joe enjoyed it, so I walked away rather than listen to them haggling. Most shops sold the same things: t-shirts, clothing accessories, inlaid jewelry boxes, and jewelry.

Water stations are for pedestrians to use.

One thing that impressed me was that there were water stations in the cities. There is one tin cup that everyone uses and is probably never washed. In the summer heat, one is too happy to get a drink to worry about germs.

The next day, after a three-hour drive to Alexandria, we saw catacombs. These are another type of burial tomb built like a tower with many stairs circling around as it wound its way down. Several large window-type openings were spaced in such a way that the mummy could be lowered by rope from one level to another.

The museum explained how mummies are created. All organs except the heart (and sometimes kidneys) were removed and pickled in a jug of alcohol-based liquid. This jug with its contents was buried with the body. The body itself was stuffed with various things and then tightly wrapped in white linen from head to toe. A funeral mask, made in the image of the dead person, was placed over the wrapped head so that the gods could recognize the person because all mummies look alike without the face mask.

From Alexandria, we went to the excellent light show that takes place at the pyramids after dark. Getting into the light show meant running a gauntlet of extremely aggressive vendors. The light show was in better English, so I appreciated the history, which was probably the same as our heavily accented guides told us. However, I found one guide's information often contradicted what another one said or what I read at the more reliable museums.

The ride back to our hotel that night was incredible. Each driver raced with other drivers, weaving back and forth in front of each other. It reminded me of Camel's driving in New York. Only here drivers had no lights, or used parking lights only, and paid no attention to lanes. We laughed a lot to pretend we were not terrified. Our driver made it to the hotel without killing anyone.

On our last touring day in Egypt, our guide insisted on taking us to see one of the children's trade schools. "The children start here at age seven and graduate at age 17," he said.

He did not tell us if it was a boarding school or a day school, but after seeing it, I guessed the answer. The only thing children learned in this school was how to weave carpets. It was a way for a merchant to get child labor at a very low cost.

The guide's reason for taking us there was to get us into the sales area where the aggressive merchant could pressure us into buying a carpet. Guides get a commission on anything the tourist buys. All four of us refused to enter the sales area and demanded to be taken immediately to the pyramids, which was listed as the final tour.

The pyramids of Giza lost some of their allure due to city housing and businesses encroaching ever closer and obnoxious vendors on foot and on camels. We drove to the biggest pyramid to find it was only open for an hour a day, and that hour was over. (The guide had to know that.) I wasn't disappointed when I learned there were 180 steps each way.

We went to the second because it was smaller and then to the third and smallest. We agreed it would be easier; it only had 80 steps and a few places where we had to crouch down to get through tunnels. Unfortunately—or perhaps fortunately—the lights went out while we were in line to buy tickets. The lights could be fixed in five minutes or five hours. People who already had tickets decided to wait, but we went on to the camel rides.

Our guide, who we no longer trusted, offered to make a deal for us, warning us that unless you knew the camel driver, it wasn't safe. Some took guests miles into the desert and refused to bring them back unless a ransom was paid. Whether it was true or a way for him to get "security" money from us, it wasn't worth the gamble. We asked to see the famous Sphinx up close before we ended our Egyptian tour.

Egypt is divided from Saudi Arabia by the Red Sea and from southern Europe and Turkey by the Mediterranean Sea. It was more primitive than Turkey. In Turkey, most farmers had a tractor or a donkey cart. We saw no tractors in Egypt and very few farms on our tour route.

In Turkey, women worked in the fields, but in Egypt farm workers were men. Egyptians had no curiosity about America and seemed intent in relieving us of as much money as possible.

In both countries, you pay to use the toilet and wash your hands. One young boy in Cairo was collecting toilet money. Joe gave him the money and then the boy said, "And one for me." Joe grabbed his money back and said, "No! None for either of you."

The boy stood in shock as Joe walked away. Everyone expected tips for everything, even things we didn't want done. The people are very poor and think Americans are millionaires. By their standards we may be. Still, it is annoying because you soon discover every time you buy anything they try to short-change you. Joe and I are not sharp on money exchange, so I only bought souvenirs when Nancy was watching.

We saw fewer women here than in Turkey, and most wore a burqa. That is the complete head covering with a dense screen in front of the eyes. With it they wore long-sleeved, ankle-length one-piece sack-like dresses (in the heat of summer) to make them all look the same and as unattractive as possible. It certainly limited their vision. Instead of a burqa, some of the young women wore a special type of head covering that concealed their hair but left their face exposed. These girls had beautiful eyes that could see where they were going.

I was ready to leave Egypt. I am glad I went but have no desire to return. I wanted to get back to the part of Africa I love. We flew to Nairobi in Kenya, where our friend, Todd Paddock, joined us for a safari that included some of Kenya and a lot of Tanzania.

In Nairobi, old cars were used as taxies. The ride to the hotel was over a two-lane road filled with potholes and streams of people *walking* in both directions on both sides of the road. It was another giant step back in time. A few lucky people had bicycles piled high with plastic crates in which they carried everything imaginable. The only other vehicles were small commercial trucks. We saw a few men walking and pulling a loaded donkey cart behind them. I guess that's what you do when you are so hungry you have eaten your donkey.

People walking to and from work all wore clean, unwrinkled clothes, just as we had found in the Republic of South Africa six years before. The ending of apartheid had no effect on Kenya. In South Africa, we saw many people walking, but not the crowds that were in Nairobi. All the shops we were taken to were enclosed in iron grill doors and windows.

We were warned not to accept candy or food from anyone we didn't know. It could be a drug to knock you out so they could steal your money. After the lies told to us in China, Turkey, and Egypt, we were skeptical, but the guys put their wallets in their front pockets, and Nancy and I carried ours in a pouch concealed under our clothes.

We passed a huge dump on both sides of the road. People were living in the cleared spaces between garbage. Some built a sort of shelter from discarded junk. I learned this had been a shantytown until someone got a match too close to home-brewed liquor, and the whole place went up in flames. There was major property damage but no serious injuries. Still, when you have very little and you lose it and have no way to replace it, it must be devastating. What a horrible way to live.

Nancy said, "They were better off when they lived as slaves. At least, they had housing, food, and medical care from the master." What price is freedom? I guess it depends on what is most important. The nomads we met in northern Thailand were happy just to be able to roam.

Our first Nairobi guide met us in the lobby, and then we walked through beautiful gardens to our rooms, which were all on one floor and next to each other. It was a day of rest, so after we took a nap, Joe and I wandered over to the restaurant. All the employees spoke English that we could easily understand. We had started back to our room after a leisurely lunch when Joe felt really sick. He was weak and out of breath, so we made many rest stops on the short distance to our room.

I thought he was tired from lack of good sleep, but looking back, I realize it was the first noticeable sign of congestive heart failure. All during this trip, he complained of being tired, having backaches and headaches but nothing we couldn't treat with Tylenol and extra naps. It was hard to diagnose Joe because he seldom complains.

The next day we went to the animal orphanage at Kenya National Park. The baby rhino showed up first followed by a couple of zebras. A guide told me things about zebras that I did not know.

"Each zebra has a different pattern of stripes just as we humans have a slightly different fingerprint. The baby stays close to its mother for several hours after birth to learn its mother's stripe pattern. When a herd is attacked, the baby must be able to locate its own mother. No other zebra will protect it. Baby zebras are brown and white. The brown gradually turns black as they grow."

After these animals were fed with a bottle, more keepers brought out four baby elephants that played with an old inner tube from a tire, drank from a bottle, and showed off for us. The guide said, "Each orphaned baby elephant has its own security blanket and one keeper who looks after him 24 hours because the keeper has replaced its mother."

That afternoon we went to see the home of the real person about whom the movie *Out of Africa* was made. It was disappointing to learn the real version. She lived alone on a coffee plantation in Africa and felt responsible for the black settlers who had taken residence on her property. She treated their illnesses and let them work planting and harvesting her coffee. In her own book, *Out of Africa,* she barely mentioned her personal life that made the movie so intriguing.

Kenya politics were extremely corrupt under Moi who came into power in 1978. In 2000, people were starving and unemployment was close to 50 per cent. Everyone claims they do not vote for him, and yet he wins every election. It is the same story as the Mugabe regime in Zimbabwe. Tourism is essential in both countries.

We had time to take a tour to a giraffe sanctuary before Todd arrived. At Giraffe Manor, the giraffe came up to the three-foot stone enclosure and stuck their heads out to be petted and to take food from our hands. One gently nuzzled my neck and face. How thrilling!

We heard the story of Betty and Jock who were responsible for this sanctuary. It began when Betty, an American, visited Kenya. She met and fell in love with Jock Leslie-Melville who lived here. After they married, they bought an old stone house on the outskirts of Nairobi, where three Maasai giraffe often visited their property.

She learned that only 130 Rothschild giraffe were left in the entire world because poachers kill them for their hides, meat, and the hair on their tails, which is made into attractive bracelets for tourists. Betty suggested, "If we can save one baby, it could live outside and just eat tree leaves. We wouldn't have to do anything!"

A giraffe is six feet tall at birth and grows one inch a week during the first year until it eventually reaches around 18 feet. It takes skill to catch even a baby giraffe. Unlike most animals, a giraffe cannot be tranquilized, so it must be lassoed from horseback. Horses are afraid of the giraffe's powerful kicking feet and speed. They found a man willing to try. He was able to capture a three-month-old female, and Betty named her Daisy.

Lassoing was only the first step. They had to walk her, kicking and fighting, to the minibus and then get her inside so she could stick her neck out the open roof. They took her to a stable where she would stay for a few days before the 225-mile trip to her permanent home in Nairobi.

After unsuccessful attempts to escape, Daisy stood in one spot staring at her capturers, ignoring the pan of milk they held at arm's length. Finally they put the milk pan on the floor and left her alone. In the morning they found her standing exactly as they had left her. She had not touched the milk.

Jock and Betty took turns sitting with the motionless giraffe, holding a pan of milk, and talking to her softly. She ignored them. After all this, would she die of thirst and starvation? Forty-eight hours passed and then, for no apparent reason, she bent her head to the pan of milk Jock was holding and began to drink. She seemed surprised to find it was milk and licked her mouth and nose with her 18-inch tongue. Then she bent down and kissed Jock. From that moment on, Jock became her mother.

This act of bonding is called "imprinting" and is turned toward one person who will return the love. It is total love mixed with the desire to please and the need to be close to someone who returns your love. Without it, a captured animal will not survive.

Daisy was so afraid that it took six people to get her into the minibus and sitting down. On the journey home, she received many stares from people who saw her head sticking out above the roof.

Jock and Betty had already built a pen (boma) for her. Nairobi is colder, so Jock hung an old tarp to protect her from the wind. It was torn and was stained with peanut butter, but Daisy fell in love with it. She rubbed against it when she was happy and hid behind it when she was frightened. It was her security blanket.

Soon Daisy was eating the trees and plants in their garden. She stayed close to both of them, but it was Jock she had bonded with. When he went to work, she mourned. When she saw his minibus coming, she ran to meet him, stuck her head through the roof and kissed him.

After Daisy was fully adjusted, Betty suggested getting another Rothschild giraffe to keep her company. This time they were able to capture a three-week-old male. They named him Marlon. At first, Daisy hated him, and when he came close, she kicked him. Betty took her chair to the boma and sat with Marlon every day. She hung a pretty curtain that became his security blanket. After many weeks, Daisy accepted him. From then on, the two were inseparable.

Daisy did not live the normal giraffe lifespan but produced many offspring. Betty and Jock were responsible for saving 23 of that original Rothschild subspecies. They were transported to national game parks where they are protected from poachers.

After Jock's death, Giraffe Manor became a tourist attraction free to city school children. Africans who may never have seen a giraffe can pet and feed them and hear Daisy's story of love.

The following day, Todd arrived and we prepared to head for the two-week safari that began and ended at Serengeti National Park.

Serengeti is a Maasai word for "endless plain" and it is home to vast herds of wildebeest, gazelle, and zebra, as well as small numbers of other wild animals. Once, right beside the road, we saw a pride of 11 lions, including five cubs that were from two to three months old. We watched one mother nurse three of them.

Each type of safari uses its own kind of sight-seeing vehicle.

It was a long way to the Kenya/Tanzania border where we filled out a lot of paper-work before we met our new driver who would be with us for the entire tour. There were five vans in our caravan for security reasons and four to six people in each van. Since we were the only five Americans, we had our own van with Hassan, who spoke good English, for our driver/guide.

On a safari, all eyes watch for animals, which, except for giraffes, tend to remain hidden in the grass. When someone in the van spots some movement, the guide takes off in that direction. Hassan was usually the one to spot something.

Often when one van stopped, others would follow and surround the spot, but if the group in that van wasn't interested, they might move on, taking the lead. This means that all day the five vans kept swapping positions in the line. The drivers stayed in contact with CB radios.

At the end of our first exciting day, we headed toward our hotel on top of a big hill. At this point, our van was in second or third place. The vans stayed a distance apart because of the dust. The first four had already climbed the hill when the fifth van appeared. In it were five Belgians who barely spoke enough English so we could communicate when we were all together for meals. Their guide spoke their language.

Besides the driver, their group consisted of a 20-year-old student traveling alone, two older men who were friends, and a young couple on their honeymoon. By the time their van reached the foot of the hill, bandits had pushed a tree across the narrow road. When the driver spotted the ambush, he started to back up. One bandit, stationed out of sight behind him, shot through the back window, shattering the glass. We think he only intended to stop them—not shoot anyone.

But the bullet ricocheted off the metal roof, flew back and grazed the side of the face and ear of the young bride. She screamed but was quickly silenced by other bandits approaching from the front who shouted orders in a strange language the driver understood. He interpreted, "They want each of you to give them $5,000 or we'll all be killed."

No one carries cash on a safari. Tourists use credit cards. Some people carry a little local cash for incidentals. Passengers and driver were ordered to get out and lie face down on the ground with their hands behind their heads.

Sunset disappears as quickly as the pulling of a window shade in Africa, and it was now pitch dark. They felt the robbers going through their pockets, and sounds suggested they were looking in purses and luggage in the van. These bandits wanted cash only. Jewelry was too easy to trace, so they did not take the wedding rings.

Those six terrified people lay face down, waiting for the sound of the first shot. When would it come? They expected to die. After a while, the rummaging noises stopped. There was long silence. Still they lay there, afraid to move until one of the older passengers stood up.

"They're gone."

People deal with fear in different ways. The bride couldn't stop crying. Her husband tried to comfort her, to no avail. Her cheek and left ear were red, swollen and painful, but fear was worse than pain. The hotel made arrangements for her and her husband to be flown home. This meant one driver would have to take them to an airport. Their honeymoon was over. I hoped sharing such an experience would bring them closer together, but it could just as easily end their marriage.

The older man, who stood up and looked, said little. He seemed to brush off the incident as part of traveling and enjoyed the rest of the trip.

His friend cussed and spouted hatred for all black people at every opportunity for the rest of the trip. It is one thing to hate, but there is no excuse for cruelty. He cussed *all blacks* in front of our African drivers. I apologized to Hassan for him. Hassan just shrugged.

It was the young man who surprised and worried us all. He was traumatized and scared. He felt the bullet whiz by his head and realized it barely missed him. Had his body shifted just a little to the left when the driver backed up, the bullet would probably have hit the back of his head. I don't know how long he saved his money and looked forward to this trip, but for him it was over. (What we didn't learn until later was that, if anyone in that van was killed, even accidently, they probably would all have died to get rid of witnesses.)

With one less driver and van, the three remaining passengers were switched to other vans. At first the boy traveled with us. It may be that he didn't speak or understand English well, but he didn't seem to talk to the two other Belgians at meals or at the lodges. Another switch was made separating the two older men. The kind man who spoke his language "adopted" the boy. The youth continued the tour because he had no choice, but he showed no interest in the animals he had come so far to see. He was waiting for it to end so he could go home. I wonder if he ever traveled far from his home again.

Although our van was never in danger, this event made a deep impression on me. I tried to put myself in the position of those involved. Had I expected to be shot at any minute, how would I have felt? I was not afraid of dying. However, had rape or torture been involved, I would be a coward.

Although I didn't want to die yet, I thought a simple execution would be a great death because it would be over quickly. Wasn't that better than a long, slow suffering death? Then I thought about the bragging rights at "show-and-tell" for my grandchildren and their offspring. I can imagine them proudly boasting, "My grandparents were shot by *bandits in Africa!*" The other kids would hold them in awe.

Every day there was something exciting to see. We saw every kind of wild animal that lives in that part of Africa and learned many facts about them. We even saw a mother and baby rhino, which are scarce. Watching lions was the most thrilling because we could get very close. One day, lions were mating a few feet from our van.

Once Hassan stopped to point out a couple of female lions stalking a group of Reed buck deer that were standing as still as statues. Then suddenly most of them ran away and a lioness ran after them. The few who didn't run were all looking in the same direction and then we saw a second female coming at a faster clip to join her sister. Still the antelope stood still. Then we saw two male lions loping along together. They climbed up on a big ant hill to rest. Hassan pulled our vehicle very close until he realized they were looking behind instead of where the action was taking place.

Lioness hides to eat her kill.

Male lions steal her kill.

Looking back, Hassan spied a female lying low in the grass, so we pulled over near her and watched her eating a gazelle she killed.

"Watch what happens next," Hassan said.

The males were focused in her direction. Suddenly they both took off running toward us and snarling. The lioness backed away from her half-eaten kill. The biggest male grabbed the carcass and ran off, leaving the second one with a few bones on the ground.

It was her kill, but she instinctively knew it wasn't worth getting hurt by bigger males. There is a time to fight and a time to back down.

At the Kenya border, we had to say good-bye to Hassan. Of the many guides, both male and female, we've had while touring the world, I liked Hassan best. He was a large, very black man with a good sense of humor and an amazing ability to spot animals hiding in tall grass or behind bushes.

We still had a day of exploring Kenya's Bushveld with a different guide. This time we were in Maasai territory. When these warriors were conquered by the English, they were forced to become pastoralists; their land was needed for white settlers who wanted farms that were considered more useful. Yet the Maasai continued to be nomads, moving their goats and cattle according to the quality of the grazing land. Over the years, their land was gradually diminished by the Kenya government taking larger and larger parcels of it.

Now the Maasai live on semi-arid lands on the Serengeti Plain where there is a constant struggle to find water and pasture for their animals. Their cattle are what they value most. They eat goat, but only eat cattle at very special celebrations. Cattle show their wealth and are given by older men to a girl's father in payment for taking her as another wife.

We visited one Maasai village, called a kraal. Before reaching it, we saw three young men walking toward the road. The guide stopped and explained these boys were starting their warriorhood, a special ritual. It is the formation of a lifetime bond that begins with a group circumcision followed by a three-month period of isolation from their village. While their wound heals, they dress in black and paint their faces in white designs. They make up their own code language, which is not shared with anyone else.

When they return to the village, there is a feast and celebration of their manhood. They cannot marry until the next age-set but can be intimate with as many *uninitiated* girls as are willing. A new age-set comes approximately every 14 years.

When the guide said we could have our picture taken with them for a few dollars, I realized they were acting out a ritual to get money from tourists. Nancy and I had our picture taken with them anyway.

The guide did *not* tell us girls also go through a circumcision ritual (cutting off the clitoris) when they reach puberty. The Maasai believe it makes a girl more valuable when she marries. (I learned about it later.) Removing the clitoris removes pleasure from sex for a female. The practice is less common now because it is against the law, but it was still being done in secret when we were there and probably still is.

Like the boys, circumcised girls are kept in isolation until their wound heals. Then one of the *older* men can take them as a bride. They may have been promised to him since birth, or it may be someone who finally has enough cattle to buy a wife. Often young boys and girls who have been intimate in youth continue to do so on the sly, but any child she has belongs to her husband, not the real father.

Even before we reached the Maasai kraal, we saw boys as young as five herding goats. When older, they herd calves and then eventually cattle because a man must be able to follow the cattle who have wandered into remote and sometimes dangerous territory.

Younger boys were often naked, but older boys had a cloth wrapped around their waist. One little boy nonchalantly lifted this skirt and peed unconcerned as our car passed him.

Young girls learn to build huts, milk cows, care for children, and do intricate beadwork to wear and to sell to tourists.

It cost fifty dollars for each car to go inside the kraal, but that was only ten dollars per person. Our kraal host had obviously received an English education. Again, I suspect this village was maintained by tourist money.

Each kraal is a small settlement of around 15 huts and is surrounded by a thorn bush fence as sharp as barbed wire. The men are responsible for tying the branches together into a strong fence to discourage wild animals from entering. As further protection, cattle and goats are brought inside the kraal at night. Younger animals are brought inside the hut and placed near the door to keep them warm and as a sacrifice if a wild animal gets inside the kraal.

Plaster for Maasai homes are cow dung mixed with urine.

Each wife builds her own hut, which takes her about seven months. This means she must begin before she is married at 12 or 13. The hut is fashioned from a curved frame built with sticks and stripped branches and then covered in a plaster made from cattle dung mixed with urine. As the mixture dries in the hot sun, it becomes as strong as cement and loses "most" of its odor.

It was very dark inside with only one tiny window up high to let light in and let smoke escape when the woman was cooking. Cow dung is used for fuel. Our host's wife led the way to a bed, which also served as chairs. There was so little space our knees touched. The bed was made of woven branches placed on the ground and cushioned with dry grasses and animal skins.

An overwhelming musty smell could have been the house, the bed, or the bodies of our host and his wife. When water is so precious, you don't waste it on bathing or washing clothes.

Our host was candid in answering questions. A man could have as many wives as he could afford. Our host has only two so far. Each wife lives in her own hut on opposite sides of the kraal to avoid jealousy. Each is responsible for caring for the herd allocated to her so she needs to bear both sons and daughters—as many as possible. Like cattle, they are her husband's wealth. I believe our host had six children at that time.

They depend on cow's milk and blood, which, mixed together, is their main source of nourishment. It replaces water, which is difficult to find. To get the blood, the husband jabs a spear into the jugular vein in the cow's neck. As the blood spills, they collect it, and when they have enough, they stop up the hole with a wad of dung and mud or with a small piece of wood carved into a small cork. The next time they need blood they use the same hole.

Our visit and picture-taking with the Maasai was a wonderful ending to our African adventure. Ron, Nancy, and Todd all flew home when we got back to Nairobi. Joe and I took another week and flew down to South Africa for a reunion with Pierre, Marjorie, and their family. We didn't know that this would be the last time we would meet in person, but we still remain in contact through e-mail.

Africa was a journey that captured my heart, and for a little while, it swept all the troubles of our real world away.

CHAPTER 20

The Last Decade (2001—2011)

THE YEAR 2001 began with high hopes. Bud and Cathie had long since taken over the running of Escapees, giving Joe and me more time to travel the country presenting seminars. In 1997 we had moved into the custom-built home our oldest son built for us. It was a house I never imagined I would own, and I was able to furnish it with brand new furniture that I picked out during dozens of shopping trips. Now we had the best of both the RV and traditional lifestyles.

This custom home was designed and built by my son, Skip. It is at Rainbow's End Escapees Park in Livingston, Texas. A pond with waterfalls and a garden are our front view. Our back view is a creek and a heavily wooded hill.

We were at the peak of our speaking careers. We gave multiple seminars at Gaylord Maxwell's "Life on Wheels" educational events in return for free advertising of Escapees. We gave seminars from Florida to California at RV shows in exchange for a booth where we could hand out club literature. And we gave several seminars at our Escapades and RV rallies.

Both of us loved speaking, but Joe always had index cards with his notes. He didn't need them except to keep him from wandering off into joke-telling. Unknowingly, that year was the end of our country-wide speaking career.

Joe needed to take longer naps and was drinking more beer to cover up his exhaustion. Yet December snuck up on us without us being aware of the warnings.

On Thursday, he complained of abdominal and chest pain so I rushed him to the local emergency room. The hospital tests determined it was *not* a heart attack. When the emergency room physician found out he was drinking beer several times a day, he decided it was an ulcer and sent us to our family physician for follow-up. The office was closed, so we went back on Friday morning. Our doctor read the hospital report and booked Joe for a GI series the first of the next week. In the meantime, Joe was to rest and take Maalox for the pain.

Joe watched TV and took long naps on Friday. He ate very little. On Saturday, he insisted the Maalox didn't help, but if it was an ulcer I didn't want to give him any aspirin-based over-the-counter pain pills. The doctor's office was closed for the weekend and someone we didn't know was taking call. I believe the pain was far worse than Joe admitted.

Sunday Cathie came to help us put up Christmas decorations; Joe stayed in bed all day. He kept taking Maalox, but that night he said the pain was worse than it had been. I should have taken him back to the emergency room, but believing the doctors who said it wasn't a heart attack, I waited for the doctor's office to open the next morning.

When I reached the nurse on Monday morning, I explained his pain was worse and Maalox didn't help. After a few minutes, she came back on the line and said to take him to the emergency room, and our doctor would meet us there.

While they were repeating the tests done previously, Cathie arrived. The report came back that he had suffered a severe heart attack and his heart muscle was badly damaged! They gave him a heavy dose of morphine, which disoriented him causing hallucinations. It took all of Cathie's and my strength to keep him in bed until he finally fell asleep.

He was transported by ambulance to a larger hospital 50 miles north and assigned to a cardiologist. He was running a fever, so the cardiologist didn't want to do an angiogram to determine the extent of the damage until his temperature came down. I camped out in a recliner chair in his room for three days. Finally, on Thursday they did the angiogram with the result that bypass surgery was needed.

"You should have brought him back to the emergency room the next day when he was having more pain," the cardiologist scolded me. "Sometimes an attack doesn't show up in tests for 24 hours."

I didn't know that, but I was already blaming myself for waiting all weekend. If I had known how much pain he had, I would have called the stand-in for our family doctor. I will always wonder if the heart damage would have been less or if the result would have been the same. Certainly he would not have suffered over an entire weekend.

People think, because you were a registered nurse, you should know more than I did. Nurses working in ICU or the emergency room would have. I had not even worked in a hospital for 30 years, yet my guilt of negligence remains to this day. The cardiologist said the damage was already done, so surgery would take place on Monday.

Both Bud and Cathie were with me now. While the doctor was telling me that, they were calling their friend with a master's degree in nursing. Alexia helped develop the heart surgery unit in a Houston hospital. She advised, "Don't let them operate there. If he requires bypass surgery, it needs to be done in Houston where they have better equipment and more experienced surgical teams. I'll make the arrangements for you." She got one of the two top bypass surgeons in the country to do his surgery!

When we told the local cardiologist of our decision, his anger was evident. "He has to go by ambulance (125 miles) and I won't sign for Medicare to pay because *we* can do the surgery here."

"We'll pay the bill ourselves. But he's going to Houston," Cathie said.

Then Bud said, "Are you going to call an ambulance or shall I?"

As soon as the ambulance left with Joe, Cathie drove my car and Bud took theirs, stopping in Livingston so I could pack a bag for myself and get personal items we both might need. By the time we got to St. Luke's Hospital, Joe was already in a hospital bed.

He was running a temperature again, but his pain was under control. They had the angiogram report but were waiting for more records. It was the weekend again, so surgery was scheduled for early Monday, but it would be done by a true expert. The surgeon agreed the main damage was already done. I stayed in his room all weekend but had to go to a hotel on Monday and stay there until he was out of ICU and in a private room again.

Triple bypass surgery took place December 18—on our 37th anniversary. Because he was 74 years old, and had had prostate cancer surgery in 1992, his recovery was slow.

On Christmas Eve, Cathie, Bud, and their youngest son came to his room with presents for us, our traditional Christmas stockings, and the cranberry juice and 7-Up cocktails that were a family tradition. It was so like them to make such a thoughtful trip to bring some Christmas joy into this non-Christmas.

Spending Christmas day in the hospital is strange. Only a skeleton crew was on duty as most beds were empty. As many patients as possible had gone home.

Joe's private room had a couch that made into a comfortable bed. I was able to stay with him as I did through all his hospitalizations whether I had a bed or not. We had many phone calls but limited visitors to family who also replenished my clothes. It was not our best Christmas, but in some ways it was the happiest. He survived the surgery, he was getting better, and we were together. Nothing else mattered.

He was home for New Year's Eve. Holding hands on the couch in our beautiful home, we talked about how much we had to be thankful for. We had both found the soul mate we dreamed of having. Our love grew deeper over the years, changing from the honeymoon stage to the helpmate stage and now to the comfortable stage of being able to finish each other's sentences.

After the mid-1980s, Joe no longer worked to subsidize Escapees. It and the CARE program were now thriving. Our house, car, and RV were paid for. Our lifestyle was simple; social security, pension payments, and Joe's wise investments provided enough money to live comfortably and afford annual trips by land and air.

We raised our children to be independent. They were leading the lives they desired, even when it was not what we would have chosen for them.

Personally, I had fulfilled all my important dreams. I had escaped deadly illnesses, written books, won many ToastMaster competitions, received standing ovations at seminars, and done more traveling than I thought possible that day when I closed the lid on my grandmother's cedar chest of *her* treasures. I brought back souvenirs of our trips, but my *treasures* were locked in my mind.

Joe also achieved the success he longed for as a child. Everyone appreciated his humor. He used jokes and funny comments to poke holes in sadness and worry so hope could filter through.

Neither of us was comfortable with me driving through Houston for doctor visits, and Joe wasn't well enough, so we went to the local cardiology team. We were assigned to a different cardiologist, but I think he was influenced by the teammate we angered by going to Houston.

He changed all the medications Joe was discharged with. We were scheduled to see him once a month but were back once or twice a week because Joe was getting weaker and felt worse every day. The cardiologist would do a test, change a medicine, and say come back in a month. We were back within a week. Joe asked for a pacemaker implant to help severe arrhythmia problems that he had had for years but were more troublesome now. I think to get rid of us, in February he referred Joe to an associate to see if a pacemaker was an option.

After checking his blood pressure, this doctor had Joe walk up and down the hall several times with an oxygen tester on his finger. While Joe was doing that the doctor looked at his chart. Then he checked Joe's blood pressure again and said, "Sir, did you know only 20 per cent of your heart muscle is alive? A pacemaker won't bring back dead heart muscle. Be thankful you are still alive and enjoy the time you have left."

During the next weeks, Joe spent much of the time napping. He was depressed and worried. Now his regular cardiologist had the same defeatist attitude as his teammate. Joe knew something wasn't right. In March he said, "I'm dying. I feel it. What can we do?"

I called the surgeon's office to make an appointment. His nurse said, "After surgery is over and the patient is discharged, there are no further visits."

"This is different," I insisted. "They discovered an aneurysm during surgery, and the doctor said to come back in three months to see about repairing it."

She made an appointment. Cathie drove us to the hospital. Joe couldn't even walk up the few steps without our help. We borrowed a wheelchair to take him the short distance to the surgeon's office.

He took one look at Joe and said, "What are you doing here?"

I reminded him of the aneurysm.

"You don't need surgery. You need a cardiologist."

"He has one." I said.

He looked at me. "Then you need a *second opinion*. He belongs in the hospital."

We asked if he could recommend someone. He had his nurse make an immediate appointment with a St. Luke's cardiologist. Less than an hour later, a new cardiologist, Dr. Younis, was arranging immediate hospital admission.

He stopped the medication and started him back on what Joe had left the hospital with and a day later made another change. Within two more days, Joe felt much better. We believe Doctor Younis saved his life. He was ready to go home, but first Dr. Younis wanted a consult with a pacemaker doctor.

After more tests, this doctor concluded Joe needed both a pacemaker and defibrillator. On April first, they implanted the combination device, and his real recovery began. Over the next few years, his defibrillator probably saved his life half a dozen times.

He remained under Dr. Younis' care for the rest of his life. I suppose when he never came back, the local cardiologist wrote him off as being dead.

Joe's heart attack changed our lives and made us rethink our travels. Our future travels by RV would be to Escapades and visiting family. We continued trips to other countries by RV, group bus tours, and ocean and river cruises.

In 2003, we went to Alaska for the third time in a "caravan" with Bud, Cathie, and our grandson, along with Glenn and Alexia Green. (Alexia is the friend who arranged for his bypass surgery. She now is Doctor Green.)

In 2004, we went to Nova Scotia with Bud and Cathie. Bud drove. Later Joe went on a cruise to the Bahamas with his friend, Todd. (I stayed home.)

Our last RV trip was in January 2005, when we went to Florida and Georgia to visit our children and grandchildren and attend an Escapees club rally. Joe's eyesight was gradually failing, and I worried that driving even a class B didn't allow for the slower reactions age was bringing.

That summer we flew to Europe for our best river cruise with Lynn and Louise Rogers. It began in Amsterdam, the capital of the Netherlands that is famous for its coffee shops where pot can be smoked legally and openly. It also has a famous red-light district where prostitutes display their wares in windows for the amusement of gawkers and seduction of customers.

The tour stopped at every major town along the three waterways we traveled down, giving us a chance to explore each town's special features with an English-speaking guide. Many famous buildings that were destroyed in World War II had been rebuilt. These tours included many cathedrals; their elegance bothered me because poor people paid for them. I can't remember how many locks we passed through, but I believe it was around 60. The trip ended in Budapest, Hungary, with the most fantastic fireworks display we'd ever seen.

In December, Joe went on another ocean cruise through the Panama Canal with 394 Escapees members. He enjoyed it but decided cruises were not enough fun without me.

Our last trip abroad was in 2006 and was a combination river, bus, and plane trip in Russia. Poor maintenance of city buildings was shocking, but most people were well dressed. Hardly anyone returned my smiles or greeting. Was it because we were Americans or was it that people were too unhappy to bother smiling at a stranger? When we asked directions, those who spoke English (young people) gave it willingly.

In late fall of 2006, it was my turn to be sick. For years I had a hiatal hernia. Several times the symptoms resembled a heart attack, which always proved wrong. When the hernia became so large it was compressing my lungs, laser surgery was recommended. Strangely enough, the surgery was scheduled for December 18, our 42nd anniversary.

Laser surgery is the easier, modern method that avoids long stays in the hospital. My hernia tore another hole within two weeks, amazing the surgeon who claimed it had never happened to him before. He wanted to do the old-fashioned open-wound surgery, but I was approaching my 80th birthday and decided to live with it.

Whether it was the anesthesia, trauma of surgery, failed surgery, or a combination of all three, I went into depression that lasted weeks in spite of antidepressant medications. During this period, Joe did chores he had never done before. He was truly a kind and considerate caregiver.

In the process of cooking, he found he liked it. He enjoyed experimenting with different combinations that used up leftovers and resulted in unusual meals that tasted good. (Some better than others). Twice a week I announced it was my turn. On those two days we ate dinner at a restaurant, and I reminded him to save enough to bring home because that was also his supper.

He tried cleaning up his cooking messes (are all men messy in the kitchen?), but his efforts didn't satisfy me. We made another pact. He cooked and I cleaned up. My job included preparing dishes for the dishwasher and putting them away later.

One day he walked into the kitchen while I was standing on tiptoes struggling to put a glass on the second shelf. I was five feet and one-half inch at my tallest, but somewhere over time I lost a couple of inches. That plus arthritis in my shoulders made it difficult to do certain things. I felt his hand taking the glass out of mine. "From now on, it is *my* job to empty the dishwasher," he said.

In 2007, shortly before my 80th birthday, I was driving to a ToastMaster meeting and didn't see the car approaching from my right. I pulled in front of him. He slammed on his brakes, blared his horn and signaled his anger with his finger as he swerved around me. It might have been a serious accident had he not reacted quickly.

This made me question myself. Should I still be driving? Yet to give up driving to weekly meetings and my "day out" would be giving up independence that I cherished. Being independent was extremely important to me now.

About a month later, I was driving to my day out, which included a ToastMaster meeting, when the sky suddenly opened, obliterating my world with blinding rain. I pulled off the freeway as soon as I saw an exit ramp. Again I did not see the car approaching from my right. This driver swerved around me with horn blaring and disappeared into the rain. There was a service station on a side street, but when I turned onto that street I misjudged the distance and wound up in a ditch.

I was sitting in the car wondering what to do when a pickup stopped. Two young men got out in that heavy rain and came to my window. When they found I was okay, they told me to do something to the car, I don't remember what, and then sit back and relax. One of the men was already attaching a tow chain, and in a few seconds they had me out of the ditch. They jumped back in their truck and disappeared before I could offer them money or even thank them. It is kindness to strangers that is so important in life and makes heartaches stop hurting.

I turned around and went home. I should not drive in this rain. Thinking I'd be gone for hours, Joe was taking a nap. When I opened the door, he knew something was wrong. After I told him what happened, I said, "I should stop driving." I was hoping he would tell me that I should still drive, but instead he said, "I'll drive you anywhere you want to go. Just ask."

I was never a *good* driver. I never actually passed a driving test but simply exchanged the bogus New Mexico license for a California license and later the California for a Texas license. I could back up if there was plenty of room, but whenever possible I drove around the block instead.

Giving up my license was traumatic, but if I caused an accident and someone was badly hurt, I would find it difficult to live with.

Joe made being dependent easier. He willingly took me anywhere I asked and read a magazine while he waited for me. Joe's life changed with his heart attack, and as collateral damage, so had mine. Our seminars ended, except one or two at Escapades. Now I could no longer compete in ToastMaster competitions or attend events because they took hours. Losing my ability to drive myself diminished my quality of life even further.

Joe's eyesight was deteriorating more with each six-month test. In the past two years, we had not used the RV—a class B Chinook. It was time to sell it to someone who could enjoy it. As we watched it slowly disappear from sight, we realized that wonderful part of our lives was over.

There were still a few places we would have liked to visit, but we didn't have the energy for long overseas flights and were no longer able to walk the distances required for side tours into towns we passed.

Joe had officially stopped tramping in 1984 but remained a union member. In the summer of 2007, we flew to California to attend the ceremony where he received his 60 -year pin as a member of IBEW (electricians union). It was a proud moment for him, and he gave a speech about the benefits of tramping that was well received.

Joe at the proudest moment of his life.

In 2008, Joe went back to what would have been his 63rd class reunion. He was embarrassed that he never finished high school and didn't like to talk about it. Like his hero, Benjamin Franklin, he educated himself, and there was not a topic that he couldn't discuss intelligently.

He did not graduate with them, but he stood proudly at the podium giving one of the most humorous speeches of his life. At the end, he reminded them the boy they thought was a dimwit and failure was the man talking to them today—a man who was both financially successful and famous in the RV world. The standing ovation is something he never forgot. That moment was the proudest of his life.

In 2009, we traveled by car and motel to the Escapade but gave just one speech together because an ulcer on my vocal cord caused by years of acid reflux had further weakened my already weak voice. Excitement now was going to town and having Joe come halfway home before he remembered I was supposed to be with him. (Yes, it happened more than once.)

Yet, in some ways our life was better than ever. We were still involved with Escapees although our responsibilities were minimal. Cathie and Bud were now training their children to take over for them. All of us were board members.

We spent more time at home enjoying pursuits we never had time for before. We became addicted to e-mail and computer research. Joe started playing computer chess with different people; I was still on the CARE Board of Directors and involved in projects with it.

My lungs had adapted to less capacity and my oxygen content was good. However, long walks or exertion affected my breathing and produced severe backaches. Joe had trouble, too, but even with only 20 per cent of live heart muscle, he was able to do things I had trouble with. Sometimes he did the shopping by himself. If I went with him, we each did part and then met so he could load my groceries into his cart. He stood in line and checked out while I sat on a bench. He loaded the groceries into the car and brought them into the house, so I could put away all but the heaviest ones. Even when he was tired, his first concern was always me.

Joe became more involved in politics. One year he would be a dedicated Democrat, and the next year, because someone had disappointed him, became just as dedicated a Republican. Whichever party he favored at election time, he donated to generously. This meant he was on both party lists when they were raising funds.

If it was his Democrat year and a Republican called, he would say, "I used to believe that way, but I changed my mind, and here is why you should, too." I think he played a game with himself to see how long he could keep the caller on the phone. Each time when he hung up, he said the same thing: "Well, that one is going to switch parties now."

He would admit he was wrong if you presented a good, solid argument. Yet when it came to politics and social issues (abortion, gay rights, etc.) he had an almost fanatical need to prove he was right. I agreed with many of his opinions, but I was embarrassed when he tried to persuade people who strongly held an opposite opinion.

When we argued about anything, he called it "having a big fight." When

I tried to end a debate saying, "It's not worth arguing about" or "Neither of us will change our mind, so let's stop arguing," he always got in the last word by saying, "You're the one who's still arguing."

I realized it was an attempt to make me start debating again. He loved to debate and didn't care which side of the issue he debated. He had a few people that he loved to argue with. If one of them said something was "black" the other would say it was "white," even if the truth was he agreed.

On Easter weekend of 2010, Janet died after a long hard battle with ovarian cancer. We had just returned from visiting her when we got the word. We knew it was coming, but you are never totally prepared. When we drove back for the funeral, Bud and Cathie met us partway. We left our car at a motel and rode the rest of the way with them. This was a difficult time for Joe. He and Janet shared a special bond.

Then it was December again. We celebrated our 46th anniversary quietly at home together. We had never exchanged presents but had we known it was our last anniversary, we'd have gone out to dinner. Three days later our granddaughter celebrated her birthday. The Carr family and Joe and I celebrated with her.

Unknowingly, she picked Joe's favorite Mexican restaurant for dinner. After his heart attack, Joe limited himself to one glass of red wine and one beer each evening. He already had his wine, so he ordered a beer at the restaurant. Everyone else was having a margarita—one of his favorite drinks. When they were served, he whispered to me, "Those sure look good."

"Why don't you order one? It won't count. This is a celebration."

Several times during the meal, he whispered, "This is the best margarita I ever had." I was glad I encouraged him to do that.

After dinner we went to the Carr's cottage so Angie could open her presents and have the strawberry pie and ice cream she preferred to traditional birthday cake. There was laughter and swapping stories and listening to Joe's same old jokes.

His theory was, if a joke was funny the first time you heard it, why wasn't it just as funny the 100th time? If you didn't laugh when he told it, he assumed you didn't get it. We knew from experience he'd keep retelling it until everybody laughed.

We all thought it was a wonderful birthday party. None of us knew it was also a wonderful good-bye party for Joe.

After we got home, I read while he played chess. About 11:30, I kissed him goodnight. He said his stomach was hurting and he was going to take a Tylenol and come to bed as soon as he finished the game he was playing. I was dozing when he slipped into bed a half-hour later. He said his stomach still hurt, and he had just taken a second Tylenol.

At two o'clock, I heard him get up and asked if he was all right. "The Tylenol didn't help," he said. "I think I'm constipated." It was a frequent problem he had.

I was still awake when I heard what sounded like vomiting. "Joe, are you okay? Do you need me?"

"No. I'm fine. Go to sleep."

Less than a minute later, he called my name in a panicky voice. By the time I got to the bathroom, he was losing consciousness.

I called 911 for an ambulance, then Bud and Cathie, who lived next door, and then left a message on our doctor's personal cell phone. (He gave me that number after Joe's heart attack to use if we ever needed him after office hours again.)

By the time Bud, Cathie, and I arrived at the hospital, Joe was conscious and our doctor was there with Joe's records. The emergency room doctor was testing for a heart attack or stroke when our doctor said, "This man has had an aortic aneurysm for over 10 years. I think it's rupturing."

They whisked him off for another test that showed the aneurysm was still intact but starting to tear and was leaking blood. They had discovered the aneurysm during his bypass surgery, but with his damaged heart no surgeon was willing to do repair surgery. It had remained stable for so many years that we had forgotten about it.

Surgery was the only hope. We wanted the surgery done at St. Luke's. A helicopter was ordered but the pilots refused to fly until the heavy fog lifted. He would have to go the 70 miles by ambulance. All three of us hugged him and expressed our love, not knowing if the aneurysm would burst before he got there.

I packed a suitcase with a change of clothes and personal items we both might need while Cathie got her car. Bud would stay behind to tend to our dogs and then follow us in his pickup. We hoped we would need two vehicles because that would mean successful surgery.

He was in a hospital bed being prepared for possible surgery, but his signed DNR stopped the surgeons from proceeding. They wanted my approval, but since Joe was conscious now, I felt it should be his choice. The pain had lessened with medication.

We went with the surgeon, who told Joe, "Without surgery you will die in a matter of hours. With surgery, there is at most a 20 per cent chance you will survive. What do you want us to do?"

"Twenty sounds better than zero. Let's do it. But doctor, if I don't make it, don't let it spoil your Christmas. I know you'll do your best."

The race against time was back on. Cathie and I barely had time to kiss him and express our love again when two operating room technicians were wheeling him down the hall. We walked behind as far as the elevator, but I don't think he knew we were there. He was telling his new audience one of his favorite hospital jokes. He must have finished the punch line because we heard all three of them laughing as the elevator door closed. (Yes, he laughed at his own jokes.)

Knowing him so well, I am positive he entertained the operating staff while they prepared him for surgery. I hope the last sound he heard was people laughing. It was the sound he loved best.

Bud arrived at the hospital within minutes after Joe went to surgery. It wasn't very long when someone took all three of us to a private room where a nurse told us Joe's heart had stopped, but the doctors were still working on him.

Then a chaplain came. I couldn't hear the conversation because I didn't have my hearing aids. It didn't matter. I remembered enough hospital procedures to know Joe was gone. The surgeon came in and confirmed his official death at 1:15 p.m. It was December 22, 2010. We were both 83 years old. I felt like I was dreaming and wished I could wake up.

Joe always thought I would die first. I wished I had. Now my quality of life was diminished even more.

The only other thing I remember about that entire day was driving home with Cathie while Bud followed in his truck. Cathie was talking—saying something important but her words were crying and I couldn't understand them, so I just kept nodding my head.

Finally, she turned to me and said, "Do you promise me?"

"Promise what? I didn't hear a word you said."

She burst out laughing, and I think I laughed, too.

The thing I miss the most about Joe is our morning and evening ritual of declaring our love and kissing. In the last few years, he stopped what he was doing several times a day, put his arms around me, and said, "Have I told you lately that I love you?"

My answer was always the same: "Not lately."

That night I went to bed thinking the world had come to an end. When I woke up the next morning, the world was still there. It was not the same world, but nobody knew it but me.

Everybody needs a security blanket whether it comes in the form of a peanut butter-stained tarp like Daisy the giraffe had or holding hands with someone you love. Joe had been my security blanket for 46 years. Now I had to find the resilience and strength to go on without him.

Epilogue
2011 & 2012

EVERYONE HANDLES DEATH differently. Joe and I frequently talked about what we would do if we were the survivor. I thought I knew, but it wasn't the way I thought it would be.

However, his actual death was exactly as he would have wanted. He went to surgery still hoping he might live, and telling jokes—the thing he most loved to do. Dying under anesthesia was the quick death he wanted and that I wanted for him. So why was it so difficult for me to accept that it was over?

That last hour when he was fighting for his life at home and in the ambulance played over and over in my mind. Gifting his body to the medical school he wanted became a miscommunication fiasco because it was the Christmas holidays. He would have been laughing at all of us when his body got "lost" for two days during the move.

His celebration of life ceremony was beautiful; he would have been amazed at the outpouring of love not just from family but from all across the country. On that day, flags were flown at half-mast at all (private) Rainbow Parks, and each park held its own ceremony.

I couldn't seem to stop talking about him to anyone who would listen. Finally, to the relief of everyone, I'm sure, I *made* myself stop. Others were going on with their lives, and I had to find a way to do the same. Dwelling on his death kept me from picking up the pieces of my broken heart and sewing them together with tears I tried not to shed in front of others. Why? I do not know.

It was strange to be alone after 46 years. I always thought I would continue living in our house and did for eight months. I rearranged furniture, emptied his drawers, gave away his clothes, and asked my grandson to erase everything on his computer. His ghost was always there.

Bud put up a memorial wall between our desks, but his ghost would not stay on the wall. It followed me wherever I went.

I remembered an old quotation but didn't know where it came from. "Don't let old dreams imprison you. Find a new dream that excites you."

I realized that living in our house was continuing to live our future together. Joe's ghost couldn't unload the dishwasher and fix the sink that would not empty. For weeks I had depended on Bud, Cathie, and our friend Todd to take me shopping and to the doctor. I knew all I had to do was make a phone call and someone would come to help me, but people have their own lives, and nobody had signed up to be my caregiver. It was time for me to make new plans for the time I have left.

I thought about buying a trailer and moving to the CARE facility. It would provide the help I needed, but it would still bind me to this park we helped build and to dreams already accomplished. Yet I could not shake the feeling people thought of me as a captain abandoning the ship in the only lifeboat and leaving those who depended on me to sink or swim.

Life taught me we each must follow our own heart, whether or not others agree. I had not started the CARE program for myself or Joe. It was for members who had no place to go when they needed assistance to remain independent.

Bud's father died 10 days before Joe and just as unexpectedly. Bud and Cathie were running back and forth between Dallas and Livingston, trying to attend to our needs and their own business as well.

In April, they said the escrow had gone through on a retirement ranch they were buying in the Texas Hill Country—a five-hour drive from Livingston. They wanted both Bud's mother and me to come and live on their ranch.

I remembered the Hill Country was where Joe's mother was born. For some reason, that brought the thought of completing a circle. There was an old building next to their ranch house. Bud planned to tear it down and build duplex apartments for his mother and me. They brought me up to spend a week and see if I thought I'd be happy there.

It was utterly peaceful. I was surrounded by small hills, trees, and pastures. Deer came to visit every evening, and there were birds everywhere. Best of all, Cathie and Bud would be there to help with anything I needed and drive me anywhere I wanted to go. Cathie *wanted* to be my caregiver and she was able to do it, whereas it would be a hardship on my other children.

Growing up in a big city with a wonderful transportation system was a far cry from living on a ranch miles from a town. Yet, I immediately felt at home. I believe it was the meaningful future Joe would have chosen for me.

Bud and Cathie buy retirement ranch in Texas Hill Country.

Bud's mother died before the duplex was started, so Bud ended up building a larger two-room cottage and bath that is perfect for me. I am surrounded by mementoes from our travels. I have been here over a year now and enjoy my independence. Yes, I wish I still had the greater independence that comes with the ability of having my own car and a driver's license. I have the next best thing: Cathie or Bud will take me anywhere I want to go.

I think back on my many accomplishments. Some are personal, like surviving tuberculosis, hepatitis, malaria, and an abusive marriage. Belief in our own ability is sometimes all we need to survive. The heartaches and disappointments molded me, gave me strength, and taught me resilience.

Some accomplishments are rewarding, like having four wonderful, loving children plus being a surrogate mother to Joe's children, who never had a *real* mother. The biggest reward was finding my soul mate and having 46 years of golden moments with him.

Some accomplishments are visible in my medals, my awards, my books. Many I could never have achieved without Joe's help and support. Together we started the Escapees organization. Its success may only be known in the RV world. The success of Escapees CARE, Inc. may only be known by those to whom it was the answer for a hopeless situation.

Life is a series of pathways, and we never know what would have happened had we chosen a different path. Because of the paths I did choose, however rough in places, my life turned out extremely well. I know my books, my columns, and my seminars have made an impact on many people's lives. People I will never meet—people I hope I inspired to fulfill *their* dream. Like Joe's erecter set, you must fulfill your dreams before it is too late.

I started to write this for our children because I realized how little I knew about my parents and their struggles to survive lack of education, a depression, a war, and cancer. Our children didn't know much more about us. We have glimpses that linger in our memories, but they don't explain *why* things happened as they did. I wished I had asked my parents more about their lives before I entered it. This book would be my gift to our children and anyone who wonders about us.

Memories dim with age, but we remember the important ones. Although conversations were recreated from my memory, the experiences I've shared here are those that impacted our lives in some meaningful way. It was time to open that closed skeleton closet where we hid our mistakes and heartaches during the years Joe and I searched for a soul mate and finally found one in each other.

Joe's legacy was his ability to make people laugh. My legacy is helping you understand our humanness. Our bad choices as well as our good choices make us who we are. How can you know us unless you know our stories?

Finally, I want you to know that both of us achieved our childhood dreams by becoming the person we *wanted* to be. I hope the same will be true for you.

Kay's ranch cottage and a different but meaningful future.

The end

CPSIA information can be obtained at www.ICGtesting.com
Printed in the USA
LVOW101935270213

321990LV00001B/1/P

9 781432 798505